'Fire? Famine? Aborigines? Who killed Australia's megafauna? Will we be next? David Horton turns the heat up on the eco-rationalists. His book will set everyone thinking again.'

—BOB BROWN, GREENS SENATOR

'Deep time and our first people's spaces are David Horton's passion. He has nailed his theses about them on the door of the third millennium, like Martin Luther, "for the purpose of eliciting truth". Horton's hoping for another reformation, an environmental one this time. I hope he starts one.'

—GREG DENING, AUTHOR OF MR BLIGH'S BAD LANGUAGE, PERFORMANCES

'This immensely readable book is a challenge to some of my cherished beliefs. Read it but hide your sacred cows first.'

—ROBYN WILLIAMS, PRESENTER, ABC RADIO'S 'THE SCIENCE SHOW'

'We are to Consider that we see this Country in the pure state of Nature, the Industry of Man has had nothing to do with any part of it and yet we find all such things as nature hath bestow'd upon it in a flourishing state. In this Extensive Country it can never be doubted but what most sorts of Grain, Fruits, Roots & c^a of every kind would flourish here were they once brought hither, planted and cultivated by the hand of the Industry, and here are Provender for more Cattle at all seasons of the year than ever can be brought into this Country.'

—LT JAMES COOK RN,
JOURNAL OF THE VOYAGE OF THE ENDEAVOUR 1768–71

THE PURE STATE OF NATURE

The fate of Australia's environment

David Horton

Routledge
Taylor & Francis Group

LONDON AND NEW YORK

First published 2000 by Allen & Unwin

Published 2020 by Routledge
2 Park Square, Milton Park, Abingdon, Oxon OX14 4RN
605 Third Avenue, New York, NY 10017

Routledge is an imprint of the Taylor & Francis Group, an informa business

Copyright © David Horton 2000

National Library of Australia
Cataloguing-in-Publication entry:

Horton, David, 1945– .
   The pure state of nature: the fate of Australia's environment.

   Includes index.
   ISBN 1 86508 107 8.

   1. Nature—Effect of human beings on—Australia—History.
   2. Land use—Australia—History. 3. Aborigines, Australian—
   Social life and customs. 4. Human ecology—Australia—History.
   5. Land settlement—Australia—History. 6. Aborigines, Australian—
   History. I. Title.

304.20994

Set in 11/13pt Bembo by DOCUPRO, Sydney

ISBN-13: 9781865081076 (pbk)

To my grandfather Charles Henry Young, my great-grandfather Robert Charles Young, and my great-great-grandfather Charles Young, all of whom would have been surprised, for different reasons, to find me writing about aspects of the Australian environment and about farming, and equally surprised, and I hope delighted, to find their names immortalised in a book.

Foreword

Why, I wondered, should I be asked to contribute a foreword to this book, written as it is by a man so erudite in Aboriginal matters as to have produced and published the first *Encyclopaedia of Aboriginal Australia*, as well as being himself an archaeologist and scholar in such matters. My only qualification, even as an anthropologist, consisted in two years of study in the very early stages of Australian anthropology—a discipline now so far advanced from its first days that it is no qualification at all. Nor was I ever in 'the field'—that field which is now crisscrossed by so many tracks that it has come to resemble the actual fields in which sheep and cattle have stamped their mark on the virgin soil of Australia.

I concluded that one possible qualification might lie in my ancestry as a descendant of four generations of pastoral exploitation of the original, dusty field. On both sides of my descent, my forebears have been breeders and exploiters of sheep, cattle and horses, those hard-hoofed animals we imported to deepen our tracks in the soil never before trampled.

Pastoralists, and convicts, were among the first to encounter at first hand the problems which are examined in this book, and some, at least, took it on themselves to begin the great argument as to the provenance, habits, social organisation and religion of those whom they were dispossessing.

So George, one of my great-great-grandfathers, who arrived in 1827 with a shipload of livestock, servants, a wife and an imminently arriving son, became acquainted in the late 1820s with tribes like the Geawegal of the lower Hunter Valley, on whose lands he squatted in days when the patronage of Governor Darling extended to him the privilege of doing so. I hope that to quote his views is not seen as too amateur a reference by an amateur.

At all events, when later Governor Gipps issued a question-naire to landholders as to how Aborigines might be persuaded to work the labouring hoe on their holdings, he much annoyed the propounders of the question by his replies: 'Two hours spent in catching an opossum or in fishing will supply them with all they want for a day, why then should they vex themselves with the drudgery of labour? They are not fools, they are not labourers at all, and for the same reason that any other gentleman is not, viz. that he can live without labour.' He characterised them as philosophers, who 'realise the philosophy that Diogenes only dreamt of, yet are not Cynics rather Gymnosophists'. This implied that they were teachers of wisdom who did not indulge in warfare or the pursuit of profit, had no need for government and so were free of fear and guilt, and lived a life of 'discipline within nature'.

This was an intelligent judgement and remains so to this day. It was a pity then that the demands of the British economy sent George and his family far afield with their flocks and herds, displacing Aboriginal occupation wherever they went in the interests of the Satanic mills of industry.

More or less contemporary with George was a convict shep-herd who became another great-grandfather of mine. His views on the nature of Aborigines have not been recorded (and very likely he was illiterate in any case, while George had had the benefit of a classical education). Probably he shared the views of other convicts and ticket-of-leave men, sent into the field to face the field's owners and their spears and boomerangs, with some-times few if any weapons capable of retaliation. To these labourers for the gentry, Aborigines were at best lazy, at worst emissaries of the devil.

These conflicting views recur in one form or another throughout our brief history, and in this book you will find plenty of them spatchcocked for your examination. From naked philosophers through to naked demons, variants of the two opposing views persist in science, religion and politics to this day. Aborigines have been accused recently of causing the death and extinction of the megafauna and, by the use of fire, of desiccating the continent, as well as of claiming land (real estate) to which they allegedly have no rights. Other moral accusations lie in the shadows ready to be brought out on occasion.

It is bad for us all to be in two minds (or more) over the status of the true owners of the land we dwell in, if not own. I commend a thorough reading of this highly informative and entertaining survey of the field as it now stands, both to clear our minds and to inform us of the present state of the theories in question.

The argument will go on, being as it is an argument between tolerance and self-interest, as long as our occupation and destruction of this land continues.

Judith Wright McKinney

Contents

Acknowledgements

As usual my thanks to my family, Vicki, Vanessa and Tanya Horton, who as always provided the emotional and mental support that is essential in completing a project like this. I am also grateful to Sheila Drummond and John Iremonger for seeing the potential in the manuscript and helping me turn it into a book. John also had the wisdom to ask Venetia Somerset to edit the manuscript, which she has done with great skill and care, and she has improved it immeasurably by overcoming my stubbornness.

A number of friends and colleagues gamely undertook the task of reading earlier drafts of the manuscript, and unhestitatingly (and without holding back from brutal honesty) provided the critical reactions needed to see where I wasn't explaining things adequately, and where I was being boring or obscure; they also encouraged me to complete and publish the book. I thank Stephanie Haygarth, Dermot Smyth, Caroline Harch, Leslie Lockwood, Bob Burne, Richard Webb and Don Drummond for efforts above and beyond the call of duty. My particular thanks to Lara McLellan, who not only provided the enthusiastic encouragement and belief in the worth of the project that every writer needs, but gave unstintingly of research assistance with the difficult and boring bits at the end.

As I completed this book I heard the very sad news that my old friend John Calaby had died. John, among his many other interests and achievements, was an early sceptic about the excessive claims about Aboriginal impact on the environment. I hadn't seen him for some time, and didn't have a chance to tell him before he died, so I can only record here my admiration and affection for John, and my gratitude for his support and encouragement over the years.

The value given to continuity is so high that they are not simply a people 'without a history': they are a people who have been able, in some sense, to 'defeat' history, to become a-historical in mood, outlook and life.

—W. E. H. STANNER, *WHITE MAN GOT NO DREAMING*, P. 38

1

'Paved with good intentions':

Theories on Aborigines and the environment

In all our stations there is a uniformity of culture only modified by the availability of different materials for manufacture . . . It is to be feared that excavation would be in vain, as everything points to the conclusion that they were an unchanging people living in an unchanging environment.

—ROBERT PULLEINE[1]

The brief story I am going to quote is the most horrifying single paragraph I have read about the Australian environment. Horrifying for what it tells us about attitudes in the bush, to the bush, and for its description of just a single small episode of the kind of casual destruction that has so badly damaged this country in 210 years. Horrifying also for its vision for the future, as economic rationalism, having irreparably damaged employment, health, education, the arts and many other areas of Australian society, turns its attention to the environment. My keyboard melts as I transcribe these words:

Five years ago, Top End farmer Bill Moon bulldozed a square kilometre of eucalypt forest to plant more rice . . . Among the

1

bigger eucalypts knocked over were about thirty nesting hollows of red-tailed black cockatoos. At the time, Moon wasn't overly concerned about the habitat of the shrieking black birds with the flashing red and yellow tail feathers. However the 30 nesting hollows knocked down at Mount Ringwood proved to be one of the most populous nesting areas ever recorded for the red-tailed black cockatoo, and now Moon has some regrets about his actions. Red-tailed blacks, it seems, may eventually bring him far more return than the rice ever did.[2]

Imagine this kind of procedure repeated thousands, perhaps tens of thousands, of times every year over most of the continent for 210 years, and wonder no more at the state of the Australian environment. Like other 'settler countries', Australia is a nation that pays homage to its settlers, seeing them as triumphing over great odds, enduring great hardship with great strength of character, forging Australian national characteristics. If the Battle of Waterloo was won on the playing fields of Eton, Gallipoli and the Kokoda Track were created on a hundred thousand bush blocks.

But a country that sees triumph over the environment, the conquest of Nature, as defining its national character, is a country in whose future the environment becomes more and more degraded and destroyed, until the settler ethos makes a country that is unfit for settlement. Instead of seeing the ancestors as heroes, the settlers should be tried retrospectively, in a kind of Environmental War Crimes Tribunal, for crimes against the environment.

Mount Ringwood is a microcosm for the uncomfortable history of European conquest, not just of the Australian continent but of the Australian environment. It raises many of the themes that are dealt with in this book: the destruction of trees, farming practices, attitudes to wildlife, attitudes to the Australian environment. Indeed it raises indirectly the question of attitudes to Australia itself. There is a very old stream of thought among some scientists, intellectuals, the media, and the public generally, that Australia is a place of the second-rate. It includes a contempt for the indigenous people and their culture, for the plants and animals, and for the landscape, and incorporates a contempt for locally produced art, literature, films etc. in comparison with those of America and Europe—the 'cultural cringe'.

Farmers would probably not see themselves as thinking of Australia as second-rate. They are likely to be of the 'my country right or wrong', 'love it or leave it', 'greatest country on earth', 'keep this our flag forever' nationalist mould. But there is a curious disjunction between the 'nation' and the 'environment'—people who see the one very clearly, and with ardent affection, tend not to see the other at all. When examined more closely, what such people tend to revere is the successful transplant of British ideas, economy, agriculture, plants and animals into Australia. Farmers revere great Merinos and Herefords, they don't revere kangaroos or possums or cockatoos. Indeed indigenous animals are likely to be regarded as vermin, and the trees and shrubs referred to as scrub, good only for clearing. Indigenous people, having failed to invent the wheel, the plough, barbed wire and the gun, are also held in contempt, and attitudes have emerged again after years of being partly hidden (at least from mainstream media) before the rise of Hanson and the Wik debate gave them a focus.

Finally, Mount Ringwood shows the emergence of a new force in environmental matters, 'economic rationalism', which could do as much damage as any of the destruction wrought by the application of British farming practices, and British attitudes, to a land that deserved better.

One aim of this book is to show the way in which past and present and future combine with history and prehistory in forming current politics and philosophy. Once an esoteric study, archaeology is a powerful tool when used or misused to promote the political agenda of various groups. This is not a new phenomenon. Nazi Germany used an invented history to romanticise the Nazi party and provide propaganda about Germany's destiny; others have done something similar. Australia has its own myths about the past—some harmless, some not so harmless. Recently, ideas about the past have been used to promote particular environmental views. This is a theme I return to several times. At a time when every month (recently, for example, the frightening tornadoes and floods in Central America and the USA) brings evidence of the ecological disaster facing the world, such ideas must be carefully examined and challenged on every point. We have enough problems without basing environmental programs on mistaken ideas about prehistory.

This book is mostly about prehistory, then, about countering some mistaken notions, which have acquired the status of orthodoxy, about the physical impact of the Aborigines on this continent. It is a plea both for understanding the original Australians and for preserving what remains of the Australian environment.

Mark Twain said of the people of India, 'It is a curious people. With them all life seems to be sacred except human life'. It could be said of Australians, 'It is a suicidal people. With them no life is sacred except human life'. We don't have to have either of these imbalances.

I live in the country, on a block partly cleared, probably last century, by intrepid pioneers. In the part that was cleared I can see the effects; in the part uncleared I see what once was; in the recovery of the land I see the potential. My small piece of land is a microcosm for the continent, like Mount Ringwood, and I will return to it again and again. In this book I aim to observe locally and think globally.

As I write this I look from my window to the east. Above the fog, I watch the sun rise over a pair of high, rounded hills. The first rays shine through distant branches, an irregular fringe of angular forms and dark-green foliage. Anyone sitting on this hill waiting for the sunrise to bring warmth to the morning would have seen the same view at any time in the last 50 000 years.

The rising sun begins to move the fog; the solid greyness begins to shift and flow. Then it starts to thin, and dimly seen shapes of trees can be seen closer at hand. This Australian landscape viewed close up is quite different to that of the distant hills. Here is a park, with a few large box and stringybark dotted across low rolling hills. As the mist thins more, other shapes emerge—in the middle distance young Hereford cattle, and closer to the house, Wiltshire Horn sheep, both animals and names resonant with their British origins. Occasionally appearing among them are what in this context seem quite alien shapes, shapes that change as the animals move, from four-legged sheep to bipedal beings mistakeable for humans. Bounding away they become identifiable as grey kangaroos, their strength and agility turning what to the sheep and cattle is a grassland divided into rectangles by barriers into a much older open landscape with no impediment to movement.

Still the sun rises, putting light directly onto the pasture, and at once the difference between a distant landscape dominated by botany, and a near landscape dominated by zoology, becomes apparent. Not some quirk of soil or aspect or rainfall, the animals have grass to eat in this place because this land was cleared. We know this because scattered through the grass, like tombstones, are the bases of trees. On most properties the process was completed long ago, and the evidence removed in a tidying-up operation where stumps were pulled and burnt, and a smooth greensward soon looked as if it had been that way forever.

Here, though, it is like a burgled house, drawers open and contents spilt, giving clear evidence of what has happened to the land. Several thousand trees were cut down, ringbarked or poisoned in a few weeks a century or more ago. It is possible that in those few weeks more trees were cut down on these few hundred acres than were cut down on the whole Australian continent before 1788.

The last hundred years is the only period in the last 50 000 when it would have been possible to sit on a bare hill in this place. In the last fifty years, the opportunities to sit on hills covered in trees have become fewer and fewer; in another fifty years they are likely to have gone.

There are other differences that I and my alter ego of 300 years ago would see as we looked around. At the bottom of the hill, ground is now bare that is unlikely ever to have been bare before. It is bare because of the light dusting of salt, an indicator of the upward movement of salty groundwater. I imagine a story:

Long long ago the land was covered by sea. The snake searched and searched and couldn't find any dry land where he could rest. So he dived down and down and hit the bottom of the sea. He opened his mouth and began to drink the salt water. He drank and drank until the level of the sea fell so far that the seabed was dry. Can I stop now? he asked the koala who had walked out onto the new dry land. No no, said the koala, the land is too salty for anything to grow, and you and I will be hungry. So the snake kept drinking. He reached further and further down into the soil. He reached down so far that he turned into a tree, and his tongue became the roots and his tail became the branches, and other good plants grew around his body, which was now the trunk. Thank you, said the koala as he climbed the tree and began to eat the leaves, but make sure you stay there so the sea never comes back again. And from that day on the Aboriginal people, who ate the other plants and hunted the animals who lived in the forest, never chopped down trees, for fear the sea would return.

With the baring of the ground comes erosion, at times into massive gullies. An erosion gully is something that my alter ego may never have seen, or seen only rarely, as a noteworthy and temporary feature of the landscape, after an unusual combination of circumstances such as drought followed by a major fire, followed by days of rain followed by drought.

Finally, not only had a land dominated by plants become a land dominated by animals, but the proportions of those animals had changed. My friend would comment on the abundance of kangaroos and parrots and birds of open space like magpies and pipits (the pipits, looking like an Australian sparrow, feed and play all round the house). He would notice the few small birds, and the virtual absence of small wallabies and possums and other small marsupials. He would probably say, noticing the lack of understorey, and the absence of hollow logs—ah, their homes are gone.

Here we sit, side by side, separated by 300 years of history and 50 000 years of differing world views. Neither of us can claim to be a conservationist; both of us want to see the land healed. How did we come to this damaged landscape, this uncertain future?

Aboriginal society and culture and religion combine to ensure that Aboriginal society and culture and religion stay the same for all time. They also combine to ensure that Aboriginal use of the environment results in that environment staying exactly as it is for all time—to 'defeat' history, in Stanner's words, by becoming 'a-historical in mood, outlook and life'. When I was studying farming I was taught that at the end of every year I should be able to look back and know that I had effected improvements, made changes, constructed objects—there should never be a year with no progress.

This is the abysmal difference in outlook that has had such tragic consequences in this land.

There are many misconceptions about the relationship between Aborigines and the environment on the one hand, and the subsequent occupation and agricultural use of the country on the other. It was a misconception that began early (the wise hand of Providence removing the megafauna to make way for cattle and sheep) and has continued (firestick farming clearing tracks and land for farmers).

For many years it was both politically correct, and anthropologically sound, to argue against the proposition that Aborigines had made no impact on the Australian environment. Politically correct because it supported the proposition, still heard today in the native title 'debate', that if you don't use it you lose it—if you owned a whole continent and you weren't using it productively, you deserved to have it taken away from you by someone who would. (Interestingly, the same proposition is often heard today from the hard Right 'develop at all costs' group, the anti-environment 'knock down all the trees' brigade, the religious 'every sperm is precious' anti-population control mob, and the anti-immigration 'if we don't rapidly build up to 100 million people the Chinese will come and take this continent away' nasties.)

There were gradual shifts in the attempts to rationalise the fact that a small group of British people had taken a whole continent away from a very large number of Aboriginal people.

The first archaeological work, an excavation within a year of landing, was an attempt to see whether Aborigines had any belief in an afterlife, and hence whether they were human or not. For many years physical anthropologists have contributed to an evolutionary belief that Aborigines were somewhere down some scale. For many years, too, archaeologists contributed to a belief that Aborigines had been on the continent only a short time, and therefore, having only just unpacked their bags, as it were, had no more right to it than the British. (Similar propositions were put in southern Africa, leading to absurdities like the depiction of the ruins of Great Zimbabwe as being recently built by slavers. In Australia there has been a comparable line of thought: from the moment that some of the art of the north and north-west was first seen it was portrayed as being too sophisticated for Aborigines, and was clearly the remains of a lost civilisation, or visitors from other continents or even outer space.)

When the great length of time of occupation was gradually unravelled and recognised, it was logically clear that occupation of the place for thousands of years (eventually growing to 50 000, probably the right figure) must give some right of ownership. This had to be countered, and it was first countered by trivialising the length of time—essentially the argument was, to paraphrase Phillip Adams's immortal words ('we haven't had thirty years of television, we have had one year of television thirty times'), that it wasn't so much 50 000 years, but one year 50 000 times. They were an 'unchanging people in an unchanging land' (although the user of this phrase didn't mean it in the way it has been misused), starting in the stone age and ending in it, starting with primitive rock art and finishing with primitive rock art—couldn't even invent the wheel, as one recent vested interest expressed it, after the Mabo case suggested that length of ownership did indeed bring some rights with it. Finally, they hadn't even had the nous to wreck the environment.

The proposition also had academic respectability. It was hard to escape the evidence that Aborigines still used stone tools and that they were hunter-gatherers. In terms of both technology and economy they were, on the face of it, equivalent to early developmental stages in Europe. The lack of local development of anything considered essential in a civilised society—the wheel, writing, houses, clothing, gunpowder, cat-o'-nine-tails—confirmed the view that, unlike the case in Europe, here development of society had been arrested at the Palaeolithic stage. The Aborigines were to the anthropologist, it was famously pointed out, what the platypus was to the biologist—a means of studying living fossils.

It was all a happy conjunction of science and politics and commonsense. Aborigines hadn't had land taken from them—they hadn't owned it (not having used it) in the first place. No, settlers had simply arrived and taken up land, a home not only among the gum trees and the kangaroos but, so many stage props, the Aborigines. The beauty of it was that, in another well-known fact, the Aborigines, like other primitive races, would simply melt away. The mechanism was 'unknown', but when up against civilisation, they just couldn't compete. It was a good example of practical Darwinism in action.

It took a brave man to stand up against all this. Orthodoxy in science, particularly when it has economic and legal consequences, is hard to argue against. Norman Tindale was a brave man, and one with a grand vision. There were three points where the orthodoxy could be attacked: you needed a mechanism to powerfully affect the environment (since it was clear, or reasonably so, that Aborigines hadn't been clearing or fencing or making roads or dams or earthworks or monuments)—fire; you needed a clear example of environmental damage—extinctions; and you needed new ways of looking at the Aboriginal economy in order to suggest that it wasn't just parasitic in nature but actually quite like farming if you squinted a bit.

Tindale put forward all these propositions. Aborigines, he argued, had caused massive change by the use of fire:

> Man, setting fire to large areas of his territory . . . probably has
> had a significant hand in the moulding of the present configur-
> ation of parts of Australia. Indeed much of the grassland of
> Australia could have been brought into being as a result of his
> exploitation. Some of the post-climax rainforests may have been
> destroyed in favour of invading sclerophyll, as the effects of the

firestick were added to the effects of changing climate in Early
Recent time . . . Perhaps it is correct to assume that man has
had such a profound effect on the distribution of forest and
grassland that true primaeval forest may be far less common in
Australia than is realised.[3]

Tindale believed Aborigines had set out deliberately on a
strategy to create grassland because people living in the grassland
areas were on the way to becoming cereal farmers—the 'Panara
Culture', he called it. Wanting more grassland, they were in a
way analogous to the early settlers, and recent developers,
who therefore (Tindale didn't make this point himself, but it
could be inferred) were no better and no worse than Aborigines.
One side effect of all this change was the extinction of the
megafauna (and Tindale thought the introduction of the dingo
had had an impact as well). So really, it was no use looking to
Aborigines for ideas on how to manage the environment; they
were just like us: change the environment for economic gain,
introduce feral animals, cause extinctions. If they were just like
us, then maybe the developers who arrived in the ship with the
spaceship name *Sirius* shouldn't really have taken the land away
from the developers who were *in situ*. The appeal of this latter
argument resulted in the adoption and expansion of the Tindale
theory throughout the Australian archaeological community, and
in an amazingly short time it became and remained the new
orthodoxy.

The new orthodoxy purportedly showed, in fact, that Aborigines had the capacity to wreck the environment had they been
so minded. The fact that they hadn't meant that they could be
claimed as the first true conservationists, and their methods hailed
as the way to proceed. If you are just living in harmony with
Nature you aren't much of a role model, but if, like us, you have
powerful tools at your disposal, and refrain from using them to
ultimate capacity, boy, are you a good example. There was a
little logical fly in the ointment—the extinction of the megafauna.
It was all very well to say on the one hand, hey look, environmental vandals, they do fit into the Australian ethos and have as
much right to the land as any bulldozer driver or chainsaw

operator. On the other hand, wiping out a great chunk of animals that look like the sort that bring tourists to Africa isn't really the act of a good citizen. The way around this has been to argue that it was really all a very long time ago (and therefore no need to say sorry) in a time when Aborigines were new to this conservation business and megafauna were new to these funny two-legged beings with spears.

Questions of Aboriginal origins have not only played a part in speculation about the causes of extinctions but have also been of considerable political significance. The idea that Aboriginal people had originated elsewhere began, probably, with the children of the first people who had arrived in Australia, by boat or raft from somewhere in south-east Asia, some 50 000 years ago. *Where did we come from, father? We come from over there, from far over the sea.* For some generations the stories would have been told and retold, the folk memory remaining alive as in New Zealand, where the epic travels of the Maori ancestors are well known.

Gradually, the ancestors took their places in the Dreaming. The tracks they had travelled became Dreaming tracks across the sea or under the sea. The fact that ancestral beings had travelled by sea is so well known in northern Australia as to be a major feature of claims for sea rights. In southern Australia, too, there were no doubts that ancestors had travelled and come from other places to the present place where a particular group resided in recent times. These are origin stories, common to all human beings in all parts of the world at all times.

Speculation about origins began from the moment of the first encounters between Aborigines and Europeans, and has continued ever since. Initial speculation was framed, of course, in biblical terms (Australia being unknown in biblical times, the origin of people living there, or the Americas, was difficult to fit into biblical accounts of the peopling of the known classical world) but later became part of evolutionary speculation. Whatever the details, it was absolutely clear that since there were no human hominid ancestors in Australia, humans had not evolved on the continent but had come from elsewhere. The scientific and linguistic tools that could be thrown at the question of where and when improved over the years, pushing back the time of arrival (or times of arrivals) and, by eliminating close relationships elsewhere, obscuring the point of origin.

Ignoring the precise details, however, Aboriginal history and the scientific method were in total agreement that Aboriginal people had arrived on the continent a long time ago, that they had come from overseas, and that they had moved around Australia in various patterns.

This agreement broke down a few years ago when archaeologists, talking about rival theories of colonisation, or linguistic relationships, or physical anthropology, suddenly began to be told by some Aboriginal people that this was all nonsense—Aboriginal people had always been in Australia. They had always been here and had not come from anywhere else. The author of a recent book[4] tells how she was apprehensive, for this reason, about discussing ideas on origin and colonisation with an old Aboriginal man. She explained her dilemma and he said, it's all right, we have stories like that too, that is, stories about origins. Indeed they do, and this impasse is therefore puzzling—how and why has it come about?

Well, it has come about for a variety of reasons, many of which will be explored in the pages to follow. In brief, an Aboriginal view that things had always been the same, and a view of time as being cyclic rather than linear, meant that there is no chronological basis to the Dreamtime, and therefore there is a sense that it is infinite in length. In addition, in a kind of escalation, while the known length of occupation of Australia (as assessed archaeologically) has been steadily increasing for the last century, the increase has met with no recognition by the dominant white culture that such incomprehensibly long occupation carries rights with it. If even 50 000 years is not enough to recognise prior ownership, then let's make it forever.

I grew up in that retrospectively golden age of the fifties and sixties. It was a time when the leaders of the country, and I, formed our personal views about the environment, or rather, were given views by society that echoed the attitudes of 150 years earlier. The Australian bush was rubbish, there to be knocked down and cleared to make way for grass and European trees and houses. If it moves shoot it, if it doesn't move chop it down. Some people leave such received wisdom and mores behind them as they grow; others retain and strengthen such attitudes, seeing in them almost a religious certainty. Religious certainty is something else best left behind. There is no problem anywhere in the world so bad that religion can't make it worse.

As a child, I too had the belief, shared with millions of my British predecessors on this continent, that Australian heaths and woodlands and forests were 'scrub', rubbish vegetation that existed only because it had not yet been cleared to make way for houses or factories or farms. It is a belief also shared by millions of my fellow citizens. It has its origins in three cultural constructs. First, the biblical saying that humans had dominion over the earth and all its contents to do with as they wished.[5] Second was the British and European mindset that did not see Australian plants and animals as being first-class, and saw Australian landscape as being distinctly second-rate. Third was a belief that this continent, like Africa and the Americas, had a potential that was wasted by its indigenous inhabitants. It needed hands of industry with a strong work ethic to buckle down and lick it into shape and make up for lost time.

In Manning Clark's first volume of autobiography he talks of going to England for the first time and the reaction of the English to him—it was like being set an exam that it was impossible (being Australian) to pass. I was reminded by the Tim Flannery theory recently (see page 17) of the long history of scientific and anthropological opinion that not just Aborigines but the plants and animals of Australia were inferior to those of Europe. In a more general sense, there is a long-held view among some biologists, too, that Southern Hemisphere plants, animals and people are inferior to those of the north. Jared Diamond is the most recent exponent of this.[6] The idea that Australia's animals are primitive, stupid, inferior, is one that has occurred at intervals over the last 200 years. The evidence for it is poor at best, but it keeps being trotted out, and has been damaging in two respects:

the mistaken idea that it could explain animal extinctions, and its use as a metaphor for the inferiority of Aborigines to Western civilisation. The plants and animals being inferior is seen as proof that Aborigines are also inferior, and the two topics are intimately entwined. For example, the lack of agriculture is seen by some as the result of animals being unsuitable or Aborigines being incapable. The whole thing is a slightly more subtle example of the colonial mentality on the one hand, and the trained cultural cringe on the other.

This belief that the potential of the continent was wasted by its indigenous inhabitants was based on the idea that the Australian landscape had been unaffected by human activity until 1788, and that Aborigines were merely 'intelligent parasites'[7] upon the land. It was a curious phrase, and the analogy it invoked with, say, lice or tapeworms was not only insulting but put in place quite the wrong image of economy and society. This was a belief with strong political approval, which it retains, because it reinforced the idea of *terra nullius*. If you didn't farm the land, you didn't deserve to keep it.

The politics of the situation was crystallised when anthropologists and archaeologists, on the side of the angels in the fight for land rights for Aboriginal people, began to see Australian hunter-gathering as a form of farming, and the Australian landscape as a managed landscape. That is, the form of management may not have been clear to European eyes, and the resulting landscapes not obviously man-made, but they were as much an artefact as the rolling parklands of, say, the Duke of Bedford.

A people who had nurtured the land in this way could no longer be ignored as rightful claimants to the land, with a title in land use as valid as that of any Queensland grazier or Victorian wheat farmer. It was the view of a new reality eventually to be confirmed by the High Court. A new orthodoxy, success crowned by Mabo, was going to be as hard to shake as the old orthodoxy, rooted in *terra nullius*, had been.

But there was another element here, and it caused some serious dilemmas and conflicts. The conservation movement had seriously developed at the same time as the land rights movement and the growth of anthropology and archaeology. It had embraced

Aboriginal environmental ethics and love of the land, and their apparent failure to damage the continent in any way in 50 000 years in contrast to the mayhem caused in just the last 200. Aborigines were, in a retrospective adoption, the first true conservationists. Anthropologists, Aborigines and conservationists were natural political allies. The logical outcome of this, which continues to be enacted now, is the involvement of Aboriginal people, people who will know best how to look after that country, as park managers and rangers.

The idea of being true conservationists is a powerful one, and people who were natural political antagonists of Aborigines, anthropologists and conservationists employed it to propound such absurdities as farmers being the only true conservationists, followed by even more bizarre claims to the title of true conservationists by cattlemen of the high country, fishermen, duck-shooters and foresters. I recently saw a news item, prefaced by the statement that farmers were conservationists, about a farmer who, having cleared almost all the trees off his land, was complaining bitterly that he was not allowed to clear the last few trees. This seemed to be part of a new push by people like the National Party and the NSW Farmers Federation to get rid of any regulations on the protection of the environment.[8] A similar campaign is being waged to eliminate the regulation of water use on the Murray–Darling river system. Just when we seem to be making progress, back we go again, down the hill.

The absurdity became less, and political alliances changed as the implications of the new views about Aboriginal manipulation of the environment became apparent. If it was true, as Norman Tindale believed, that there wasn't a single habitat in Australia that hadn't been created by Aborigines, then Rhys Jones's question[9] as to what we wanted to preserve, the environment of 200 years ago, or one of, say, 70 000 years ago, was a good one. If the answer, as Jones pointed out, was 200 years ago, then we would need to use fires as Aborigines had done. That is, if the Australian environment was an artefact of human behaviour, there was no reason why it couldn't be manipulated by farmers, cattlemen and foresters. The campaign to change public opinion has focused on apparent changes to the environment caused by cessation of burning, the risk to property caused by bushfires (and the need for control burning), changes to vegetation (and the need for forestry practices to rectify things). The concept that Aboriginal burning practices may have kept woodlands relatively open, and that therefore a particular spot of ground may (or may not) have more trees now if it has not been burnt than it would have had twenty years ago, has been extended into the ludicrous suggestion that there are more trees in Australia now than there were 200 years ago.

If the extinctions of large fauna—one of Tindale's theses—were caused by humans, then a few more unavoidable extinctions caused by modern activities are just too bad. Extinctions happen, and if Aboriginal use of the environment resulted in extinctions then who are we to try and stop them? In fact, it has been argued, anything nearly extinct is not worth saving, and Nature should be allowed to run its course (perhaps with a pillow to soothe the dying race). Furthermore, if Aboriginal action caused extinctions, then it irrevocably, and very early on, set in train events that mean that not only should we follow Aboriginal practice in manipulating the environment by fire (and operations of comparable effect, like thinning trees and grazing), but it is *essential* that we do so. Those practices, it has been argued, were necessary to maintain in balance an environment put out of balance, inadvertently, by a big mistake that Aborigines made 50 000 years ago. It is a marvellous concept of original environmental sin in a Garden of Eden, beside which the ravages of the last 200 years are just a surface scratch.

With this final piece in place, the game set in motion by the opening gambit of Tindale has apparently ended in checkmate for the conservationists. Anthropologists have become the political friends of farmers and foresters and big business in an unholy alliance that couldn't have been contemplated thirty years ago. Economic rationalism has little place in society and should have no place in managing the environment, but this has been the outcome of the original good intentions of Tindale and many others subsequently.

The most recent exponent of the views of Tindale and the others has been Tim Flannery in *The Future Eaters*. More worrying than the actual theory (theories are fine; they are how science proceeds, as new ones are formed and old ones discarded) is the reaction to it among the media and the general public. Suddenly, it seems, theory became fact, and there were obviously many people out there for whom someone purporting to know the truth about the past would be seen as a hero. Indeed, the reaction to Flannery is more akin to a religious movement where a charismatic leader speaks and his words are beyond question.[10]

Flannery's theory seems to be this. Australian animals are dumb, therefore when humans arrived they were easy to kill, so easy to kill that many went extinct very rapidly; because they were no longer eating grass, the grass grew long and there were many fires, so many in fact that the vegetation changed; when the vegetation patterns changed, this caused a change in climate; Aborigines, in order to deal with all this, learnt to use fire themselves to 'manage' the environment; therefore, we too should use fire to manage the environment.

In this book I will show that every aspect of this theory, and the chain of 'logic' that connects the pieces, is wrong. Aborigines didn't cause extinctions, and these didn't in any case happen soon after human arrival; extinctions don't cause changes in length of grass, so fire wasn't the result of extinctions; fire doesn't cause vegetation change, vegetation change causes fire change; vegetation change doesn't cause climate change, but the reverse is true; Aborigines did, to some extent, attempt to manage the environment by the use of fire, but they fitted into the natural Australian fire regime, and their use of fire has had little, if any, effect on vegetation; we shouldn't use 'control burning' to manage the environment, though we may wish to use it selectively to protect property. But wherever it is used, it is not good for the environment.

My counter-theories to all this form the chief content of this book. They can be supported by a true understanding of the archaeological evidence. This book (and Chapter 2 in particular) looks at the way archaeology works, at the things we can learn from the past, and at some possible interpretations of the evidence that have been found. It also presents views about what useful meanings we can draw from the past to help us through what is a very uncertain future. It does so, I hope, with a healthy dose of scepticism about how much it is possible to write history in the sense of knowing what 'really happened'. While I don't totally subscribe to Henry Ford's view that 'all history is bunk', it is certainly a better starting point than the belief some people hold that they know the truth about the past and act accordingly.

One of my strongest messages is to beware of gurus, whether religious or political or scientific. No one has all the answers. No one is worth following with a mind in which all disbelief has been suspended. If we fail to question everything, we will soon forget how to question anything.

Just a few years ago, my subject matter—Pleistocene extinctions, fire, Tasmanian prehistory, origins of agriculture—belonged in the realm of esoteric academic debate. Conversely, subjects such as land rights, conservation, drug laws, control burning, exploitation of wildlife and gun control were matters for political debate. In recent times, all of these topics and more have become blended into a political discourse.

The nature of the protagonists in this discourse has also changed. Until the 1990s you could predict which players would line up to contribute to any discussion, and, in advance, what they would say. On matters related to the environment, ecologists, other biologists, archaeologists, conservationists, left-wing politicians and Aborigines would argue for conservation of the environment, protection by government legislative action and preservation of wilderness, and against mining, farming and forestry activities involving development that threatened sites and habitats and endangered species, hunting and other exploitation of wildlife, and so on. On the opposing side would be conservative and populist politicians, foresters, miners, farmers, talkback hosts and big business.

In summary, the Left included scientists and believed in conservation and protection, while the Right included businesspeople who believed in development and exploitation and had no time for conservation and protection. In the 1990s a new phenomenon has arisen: right-wing scientists who promote the idea that you can conserve and protect the environment by developing and exploiting it. This proposition, that you can have your cake and eat it, like the idea that a rich man can enter the kingdom of heaven, is immensely appealing to the Right. It is an appeal given added relish by its apparent scientific respectability. Look, the Right can say, scientists agree with us, this is not just greed and ideology. It is a similar phenomenon to the one in which religions that see the need to 'disprove' the theory of evolution find scientists willing to argue that impossible case. Or religions that have an ideological objection to birth control finding population biologists and economists to argue that the population of the world must be allowed to grow and grow and grow. Rude things are said about lawyers, but these days you can find someone to argue in favour of any ludicrous or dangerous proposition.

It is in the development of these ideas about exploiting the environment to save it that the amalgamation of scientific and political discourse has taken place, and the relationship between what were formerly unrelated topics and attitudes has become clear. This book explores that relationship.

The new 'economic rationalist' approach to the environment can only be sustained, I believe, by a misreading and misunderstanding of history. If this misunderstanding continues, the support of powerful political, economic and media forces will increasingly ensure that this approach holds sway. Unfortunately it is an approach that will cause enormous environmental damage, which will be increasingly impossible to correct or reverse. This book also aims to combat the economic rationalist approach to the environment.

2

'An unchanging people in an unchanging land':

Archaeology and the past

A people without history
Is not redeemed from time, for history is a pattern
of timeless moments.
—T. S. ELIOT, 'Little Gidding', *Four Quartets, Collected Poems 1909–1962*

'And where are we going today, sir?' 'I'm an archaeologist, off
to a dig.' 'An archaeologist, eh? Digging up those dinosaurs, I
suppose.' It is a common mistake, and not just by taxi drivers.
A reflection partly of the appeal of dinosaurs in popular culture,
and partly of the inability of the general public to have a clear
feeling for deep time and its sequence.

The disappearance of the Beaumont children thirty years ago
in Adelaide is one of the great unsolved mysteries of Australia.
In spite of reports from real or apparent witnesses, it is impossible
to know what happened. Someone said she saw the children in
a car. Did she? Did she see anything? If she did, what was it? If
it was a car with three children in it, were they the Beaumont
children or some other family who went on their way oblivious
to the mistaken identification? And even if this sighting was
correct, what happened next? There is no way of knowing.

A year or so ago a clairvoyant claimed to know where the Beaumont bodies were buried. The concrete floor of a warehouse was dug up and the ground excavated, fruitlessly, following his instructions. The archaeologist is in a similar position to the clairvoyant, going down to the excavation to check out theories.

Whenever we try to investigate history, even history we have been personally involved in, a recounting even of a sequence of events, let alone an explanation of how and why they occurred, is near impossible. People's perceptions differ, just like any eyewitness reports at the scene of an event, and one's memory fades to the point where one's own recollections quickly become suspect. What chance, then, of writing bigger history? And what chance of writing prehistory, where even the deceptive comfort of written documents is not available?

So why do archaeology? It is a question often asked by visitors to a site, watching with amazement as you proceed to slowly dig several cubic metres of dirt with a small trowel and then sieve the result. It is indeed a question I often used to ask myself while sweltering in a trench where the temperature was 40°C, or breaking ice in a bucket of water to wash my hands in winter. Well, 'archaeologists conquer history', I am tempted to reply, keeping in mind a goldmine in T-shirts and bumper stickers.

Then why do archaeology in Australia? This variant of the question is posed by those who can indeed see the point of excavation in places like the Middle East, where one is discovering buildings or whole cities, and the smaller finds include gold ornaments, pottery and statues. Hunting for treasure is one of the pursuits that led to the development of archaeology, and the treasure-hunter image has been maintained in popular films and books, and in the press. The general public can see little evidence of treasure being found in Australia, and can therefore see little point in archaeological effort.

Rhys Jones made a comparison between Australia and New Zealand (and much of the rest of the world), noting that the effort Maoris had put into building great forts suggested that 'every society expends its excess energy in culturally desired ways—war or religion—the one invested in altering the faces of ridges and hills, which so impress our own utilitarian eyes, the other spent on bright, full-moon nights, reaffirming man's unity with Nature'.[1]

Many of the early explorers of the continent, and later explorers of the past and the environment, have seen Aborigines as 'noble savages', at one with the environment, not altering or damaging it and therefore, in the 'original affluent society', having time to invest 'into the realms of the ego, the mind and the soul'.[2] No monuments or public buildings or forts, then, and no farms. Even the tool kit had to be limited because these people moved seasonally and couldn't carry much with them. Possessions had to be portable and often multipurpose, or made of readily obtainable materials, such as stone, wood, bone and shell. Doing archaeological research in Australia is a subtle task, seeking slight traces of these materials in ephemeral campsites and using all the aids of modern technology and techniques to reconstruct the lifestyles of their occupants. Even with such aids, the reconstruction is largely limited to economic and technological matters. Except for art, finding evidence of the cultural and religious lives of hunter-gatherers in the past is difficult because of the perishable nature of so much of the materials used in ceremonies and performances. Archaeology is nevertheless intrinsically appealing because you are revealing (or hoping to reveal) the unknown.

I recently visited my old high school, in my time brand-new on a freshly cleared piece of ground. I looked, some forty years later, at the route I and my friends took between the school and the railway station. What had been a flat path was now worn down in a groove perhaps a metre deep. I was reminded of my awe at Chartres Cathedral, seeing paving stones worn down by 900 years of pilgrims' feet. Such sudden glimpses of the reality of time are rare in everyday life, but excavation also gives you an uncanny sense of time past. You are constantly aware, with a feeling of awe, that the artefact you have just found was last seen, handled and lost by people thousands of years ago. It is also true that the surface you have exposed with a trowel was then the ground surface on which people in that distant time were sitting, performing, thinking, being. In descending a ladder into the trench each day you are propelling yourself back into time more surely than Dr Who's Tardis, and each day the ladder reaches a little further back.

Archaeological work, and the interpretation of results from it, is an intellectually challenging exercise. We are not digging up *things* but *people*. The objects we find are not so much of interest in themselves but for what they tell us about the people who made them and used them. An excavation provides three different kinds of information, all equally important in writing prehistory. The first is the soil containing the objects. Determining which are discrete levels, what shape they are, and what materials they are made of are all important in understanding the history of the site, what sort of physical environment people inhabited, and the climatic conditions in which they lived.

The second kind of information gathered by the archaeologist is the arrangement of the objects found—the relationship between them. Knowing that a hundred artefacts were obtained from a site tells us very little. Knowing that ninety-nine came from a layer of clay 10 000 years old and one from a 1000-year-old sand layer above this tells us a great deal. Knowing that most of the ninety-nine were found in one corner of the trench next to a charcoal-filled depression with numerous burnt bones also provides a great deal of information. The relationship of objects, in conjunction with a study of the layers, provides information about the sequence of events in a site and possibly about the way people lived at different periods. For example, you may learn how the campsites were organised, or what activities were carried out in different areas.

Finally, the objects themselves provide considerable information about how they were made and what they were used for, and to some extent how old they are. It may be possible to find out where the stone was obtained from (and therefore learn about either local mining activities or trade from distant places). Bones can give information about hunting techniques, diet, time of year the site was occupied, the environment at different times, and even the likely age of the site (if extinct species are present). Bones, as well as charcoal, wood and shell, can also be used to provide very accurate dates, though these must be carefully checked against the section, and a series of dates is needed from different depths to be certain of a sequence.

The intellectual challenge while excavating is to absorb as much of this information as possible while it is being obtained and use it to develop ideas about the structure and history of the site. These ideas in turn are used to make predictions about what is likely to be found next (wrong predictions enable modification of the ideas; correct ones help to confirm them) and to plan the excavation. We constantly need to make guesses about what we will see next, in order to know how and where to dig next. The more we excavate a single site, the better those predictions become until, ideally, we reach a point at which predictions are accurate enough for our purposes, and then we stop digging. A very simple site will take very little excavation before it is understood; a very complex site may take a great deal.

These techniques are used to try to determine the history of a particular site: what did people do there and when were they doing it are the two great questions of prehistory (rather like the questions about Richard Nixon—how much did he know and when did he know it?). After the fieldwork, a great deal of analysis needs to be done to fit all the clues together into a coherent story.

Even the best of sites, however, contain only a small part of the evidence relating to the activities of the people who lived there. A great deal of material can never be preserved and some is rarely preserved. Objects made from plants or the soft parts of animals are seldom found, so that in Australia we generally have no archaeological record of houses, rugs, baskets, spears, cloaks, plant food, string, waterbags, boomerangs, musical instruments, spear-throwers, nets, shields, carvings and so on. This extraordinary bias in what is preserved in the soil makes for great difficulty in interpretation. Even material of a kind that can normally be preserved may be destroyed or removed from the site at various times under particular circumstances (for example, by flooding or soil acidity).

Another problem is that no Australian site was used for more than a part of the yearly or even daily activities of the people who lived there. Sites were used in particular seasons to exploit resources or make use of shelter. During a typical day, people would move considerable distances from a camp in order to hunt or collect materials or carry out ceremonial functions and so forth, and each activity could leave behind a small site with particular kinds of artefacts and other remains.

It is also extremely unlikely that any one site will have a continuous record of the occupation of a particular area through time. Rock shelters that are occupied in one period may be unavailable in another because of rock falls; new shelters may be formed and occupied; the environment may be rich enough to support long-stay camps at one time, but too poor to support more than short visits at others; changes in climate may blow away sand-dunes or wash away campsites; changes in social structures may change the size or position of campsites.

All these difficulties make the interpretation of archaeological evidence and the subsequent writing of prehistory of an area a challenging and exciting problem. The challenge and the excitement become even greater when we attempt to put together evidence from the whole of Australia to try to write the prehistory of a continent. In doing this, we are looking for general changes and trends, attempting to understand how people coped with living on this unique continent, and how Aboriginal civilisation gained its unique qualities.

Archaeologists are therefore trying to conquer history in another sense, pushing further and further back into the past, and seeing the events of a lost time with increasing clarity. But the past defends its secrets stubbornly in a constant rearguard action. Being locked into the present as we are is like missing a sense—a sixth 'sense of the past' would be as useful as any of the other extra senses for which claims are made, and unfortunately is just as unlikely to exist.

The questions posed by Australian archaeology are big ones, and increasingly the attempted answers have implications well beyond the scientific community and into the political, social and environmental worlds. Where did the Aboriginal people come from, who were they, and when did they arrive? How did they cope with the radically changing environments from 50 000 years ago to the present? How did population and social structure change over that period? How did people respond to all these changes by adapting their economy and material culture? Did they try to modify the environment in a serious way, and indeed were some of the changes the result of human activities? What kind of regional variation was there—how did people respond to the radically different environments and histories of Tasmania, Arnhem Land, the desert and the eastern highlands? Although these questions are interrelated, this book concentrates on the environmental issues in Australian prehistory.

The fact that there are no definitive answers to such questions, and never will be, adds to the enjoyment and challenge of the analysis. As long as we realise that all we have are hypotheses, and that these continue to change over time as new evidence emerges, and as fashions in the associated political beliefs change, then the hypotheses current at a particular time can be seen as part of the intellectual challenge. The process only becomes dangerous if hypotheses are perceived as truth, implications are derived from them, and politicians and the public become convinced that certain agendas must be followed. Archaeology doesn't always conquer history—history is elusive, always one step ahead.

Australian archaeology did not have happy beginnings. A perception that few rare or beautiful objects are to be found in Australia began with the first fleeting observations by people like William Dampier. Observations were brief, by untrained observers, and often in places or circumstances where little could be observed. From 1788 there were people in a position to make more detailed observations, but the imperatives of survival and development of the colony left little room for detailed observation. As time went by, relations soured in the atmosphere of mutual suspicion and murder, and not only were observations not made, but there was every reason to portray people whose land you were taking, and who you were killing, as primitive savages. Finally, southern Australia saw very rapid human alteration of the environment for farming and grazing, and these changes hid or destroyed many Aboriginal sites and structures.

The popular view of what Aboriginal societies were like has, as a result, been largely based on those of the central desert and northern tropics. Recent archaeological and ethnographic work is gradually providing a picture of the very different southern societies, revealing aspects of Aboriginal culture that were previously little known. In parts of the south, it now appears, large sedentary populations occupied permanent stone or mud houses. The people wore extensive clothing with many decorations. Their economies included large areas of grain being harvested and crops being stored, and with eels and fish controlled and harvested. They carried out large-scale mining operations, traded over vast distances, constructed large stone arrangements and had major ceremonial gatherings.[3] This is certainly not the perception of Aboriginal society painted by Morgan, Blainey and Flannery.

It is clear that archaeology in Australia can be of interest to its practitioners, but why should the public be interested in this research, and indeed why should they support it with funds?

There are two major reasons why a knowledge of Australian prehistory is important to Australian people today and in the future (there is additionally one aspect of Australian prehistory, the cause of megafaunal extinctions, which is of crucial importance to the future of the world and to the argument of this book about preserving what remains of the Australian environment). For a long time, it was possible for white Australians to think of Aboriginal Australians as being relatively recent arrivals on this continent, and as a result to suggest, or believe, that they had very little more claim to the continent than the British did. A variation of this, designed to cope with the reality of 50 000 years of occupation, has picked up on the hypothesis that different groups of Aborigines arrived at different times. The most recent group, according to this theory, had displaced the older people, thus validating their own subsequent displacement by the British. Or, as former senator Bill O'Chee put it, 'tough luck'. Another claim we have already seen is that Aborigines really hadn't made good use of their time here. They were representatives of Palaeolithic man, unchanged, they hadn't used the land productively, didn't have a written language or use metals, and hadn't even invented the wheel (as Hugh Morgan put it)—they were uncivilised, primitive savages.

Seeing 'civilisation' as something that only Europeans had has a long and continuing history. 'Wogs', of course, began at Calais, and at various times many civilisations, including the Chinese, the West Africans, the Arabs, the Incas, the Albanians and so on, have been considered worthless. The recent upsurge of the Ku Klux Klan and related groups in Australia brings new claims that only the 'white race' has invented anything. Where this is not simply unhealthy nationalism (surely a tautology), it has been part of the process of justifying conquest and colonisation, and ethnic cleansing and genocide, by dehumanising the inhabitants of an area.

Archaeological work over the last few years, however, has now made it clear that the British invasion of Australia brought Captain Phillip and his group into contact and conflict with one of the great civilisations of the world.[4] The consequences were comparable with those that occurred when Europeans contacted other civilisations in the Americas, Africa and Asia, which had also developed in isolation. Had the early British colonists been aware of the complexity and the achievements of Australian Aboriginal civilisation, it might, just possibly, have improved relationships between the two societies. These might have been based on mutual respect, and much of the course of subsequent Australian history might have been different. Making the knowledge now available more accessible and widely known just might improve present and future attitudes of non-Aboriginal people towards the indigenous inhabitants of this country, and perhaps make those inhabitants a model we could usefully emulate.

The contrast has often been noted between the sufferings of early European explorers in country they saw as harsh and barren, and the ability of these areas to support large groups of Aborigines. The most notable example is that of Burke and Wills, starving to death within sight and sound of an Aboriginal tribe who had lived there for tens of thousands of years. The problem was that, except for the wonderful Ludwig Leichhardt, the explorers operated by trying to transport European conditions with them. Long trains of mules, bullocks, horses or camels carried tea, flour and sugar, while sheep, goats and cattle were driven along behind to provide fresh meat. It was like the exploration of space, where the capsules have to provide totally for the needs of the astronauts for the whole trip. Leichhardt knew the futility of this approach and, travelling light, lived off the land as much as he could by observing and copying Aboriginal use of plants, animals and waterholes.

The explorers were only following the same procedure as the colony as a whole. The British had brought a European lifestyle, agriculture and artefacts to this country, attempting to make the country conform to them rather than they conforming to the mores of a new land. It was something like the terraforming process that colonists on Mars might undertake, where a biosphere is gradually expanded outwards. The massive soil erosion, salinity problems, and the loss of plant and animal species, are some of the damage this attitude has caused in just over 200 years in a land that had supported Aboriginal people comfortably, without damage, for some 50 000 years.

Aboriginal people had realised, perhaps instinctively, perhaps as a result of some bitter experience, that this country had to be handled like fine china. We have been like bulls in the china shop. If we are to be true Australians, and not simply transplanted Europeans in an alien land, we must learn about Aboriginal civilisation, history, and relationship with the environment. These people are so intimately linked with the land that understanding Australia is inextricably linked with understanding them. Fundamental to this understanding is the way we view the past and the evidence from the past—and the different way Aborigines, whose past it is, view this past.

John Hunter, captain of the *Sirius*, went in April 1788 to survey Middle Harbour and by chance conducted the first archaeological excavation in Australia. On 22 April he and a companion, William Bradley, landed on a point and

> found the earth thrown up in the manner of a grave, which we turned up and found the ashes of some deceased person and by the burnt wood laying near it we suppose it to have been consumed on that spot. Some pieces of bone were found not quite consumed but too much to know what part of the body

they belonged to. From the greater quantity of ashes at one end than the other, I suppose the body to have been laid at length before the fire is applied to the pile and conclude that they dispose of the dead in the same manner. Saw very few of the natives.[5]

What were they up to? Well, the last sentence gives it away. It was very hard to observe the Aborigines, especially after they had learned what the muskets could do (an early argument for gun control). But there was enormous interest in the original inhabitants of Sydney Cove for two reasons. First, they were potentially and actually a threat to the small colony. Spears were thrown and people killed, and although shots were fired in response and other people killed, it was generally recognised that the spear-throwing was usually in retaliation. Arthur Phillip was worried that the Aborigines might set fire to crops or food stores, and this would have been the end of the colony. So they needed to know what they were up against. There was also an interest in the validity of the displacement of these people by the colonists.

Terra nullius is not a statement of stupidity—everyone knew there were Aborigines there—but a judgement of level of humanity, level of civilisation, use of land, and, by implication, rights to the land. The economic system didn't help much—everyone had to eat—but it didn't at first sight look like a system that involved sophisticated land use or land ownership. James Cook said they 'have no fixed habitations but move from place to place like Wild Beasts in search of food'. So really religion was going to have to be the touchstone.

This was much more difficult to come to grips with. The British could see no evidence of religion as they understood it (no churches and no priests, at least in a form they could recognise) and they were in no position to understand Aboriginal ceremonial life even if they could have observed it. Indeed the question of whether Aborigines 'had religion' or not was to be fiercely debated for well over a century. The only handle that could be got on this question in 1788 was through burials. Humans buried their dead. Humans who buried their dead with some ceremony (e.g. cremation) were at a higher level than those who simply dug holes in the ground (though this was an awkward piece of logic, since the British buried their dead in holes in the ground, and would not be permitted to cremate for another century). Furthermore, a careful treatment of the dead implied some level of belief in an afterlife, thus providing a basis, though perhaps a rudimentary one, it was thought, for religious belief.

So Hunter was using archaeology to illuminate the present, not the past, and to settle political questions, and in both endeavours he was anticipating future uses of archaeology. For the Aborigines, archaeology was an alien notion. The idea that people would dig up a grave would be not merely sacrilegious and insulting, but it would be as incomprehensible as the actions of those same people in chopping down trees. The reasons for this are intricately linked with Aboriginal culture and views of the nature of the past.

The Dreaming isn't an earlier period but a different universe. There is no past because each individual is locked into a set of generations in exactly the same pattern as all previous generations have been. An individual has parents and grandparents and will have children and grandchildren, and so has everyone else who ever existed. The view is like that of a bounded universe, where you can see as far as you like but you always end up seeing the back of your own head. The pattern was (and is) little different in the peasant societies of Western Europe—a sense of great-grandparenthood is one for scholars and the rich and noble, not the ordinary person. You couldn't have a chronology like that of James Ussher, Archbishop of Armagh, who in 1650 published the date 4004 BC for the creation of the earth, because in the Aboriginal world view there is no sense of a human life or generation representing a certain number of years. Each generation is rather a unit, and the units always add up to zero (if you represent yourself as 0, children are +1, parents are −1, grandchildren are +2 and grandparents −2). There is no accumulated 'value' that can provide a chronology.

But how would archaeology, which at least implies a time-depth, fit with Aboriginal views of the past? 'We have been here forever' is the saying that is normally seen as conflicting with archaeological interpretation of stratigraphy and dates, but it is a statement that doesn't actually make any sense in the traditional view of the past, which is essentially a belief that 'we exist therefore we exist', like the steady-state cosmology compared to the big-bang cosmology. Aborigines were not saying we have been here a 'long time', but rather that there is no sense in considering 'origins'—human beings exist because human beings exist; matter is created continuously in the presence of other matter which has always been there. To have a big-bang model of history requires a sense of discontinuity, a knowledge of the movement of peoples, of wars, of land conquered, of voyages undertaken, of languages changed, of monuments built by earlier generations personally unknown.

When Robert Pulleine used the famous phrase 'an unchanging people in an unchanging land' in describing Aborigines,[6] he was unwittingly to incur the scorn of later archaeologists. The phrase was quoted over and over again as archaeologists, desperate to overturn what they saw as not only a grossly mistaken view but one which was derogatory of Aborigines, scrambled over each other to document the extremely changing nature of both Australia's environment and people. The idea that being 'unchanging' was derogatory seemed logical at the time, but in retrospect is somewhat puzzling.

At its simplest level, the response was an antidote to the idea of Aborigines being a kind of living fossil. Archaeologists were saying: look, Aborigines have changed a great deal since the time of their first arrival 50 000 years ago. But it is hard to see in what way change is a measure of virtue. How much change, how much stability, is the sign of civilisation? Is the maintenance of royalty in Britain and the Emperor in Japan a sign of stagnation? Is the invention of endlessly new consumer products the sign of civilisation? Is degradation of land a sign of civilisation? Is the maintenance of sectarian and ethnic hatred through hundreds of years in Northern Ireland and Bosnia a sign of civilisation?

Is there a sense of an attitude that real civilisations do endlessly change, and that stability in Aboriginal society is a sign of stagnation? Certainly this is the argument that has been used for Tasmanian Aborigines, and by comparison other Aborigines are seen as being more advanced because they have changed. But a different measure of worth might be to value stability, to feel that a society that has adapted itself to the environment and maintained itself (and the environment) indefinitely is doing a pretty good job.

There is another sense in which the rush to discredit the Pulleine line by archaeologists runs contrary to Aboriginal interests. Whatever the rest of the world and Hugh Morgan think, Aboriginal people themselves believe, and have always believed, that they are an unchanging people in an unchanging land. To argue against that belief, in order to counteract the beliefs of those who think that lack of change in some things means you remain primitive in all things, is to do Aboriginal people a disservice.

Aborigines not only believe that they have always been in Australia, but that they have always been in the particular place that their known ancestors came from. They also say that their customs, beliefs, practices and culture have all been handed down unchanged. The Dreamtime is no more a statement of evolutionary development than is the account of human origins seen in the Book of Genesis. The Dreamtime is a timeless period in which things became the way they are, and then remained so for all time.

There have been three main cosmological models about the history and future of the universe, and these (perhaps curiously, perhaps not) match up to various religious beliefs about the same topic. Given that the universe is expanding, you can believe either that it came into existence at a particular point 'in time', or that there is a constant creation of new matter occurring, generated as a result of the existing matter being present. In the latter view, the universe always looks approximately the same, and has existed forever. If you believe the universe came into being at a particular moment, and is expanding, then either it just keeps on expanding to a point where every piece of matter is infinitely remote from every other piece, or it expands to a point then collapses back on itself, eventually leading to a new 'big bang' and a new expanding universe. This cycling could continue infinitely.

Aboriginal world views are often thought of as cyclical (thinking of time in cycles), but in fact more closely resemble the steady-state model. The world has always existed because it always existed; Aboriginal people have always existed because Aboriginal people have always existed. Sure there is a period of creation in the Dreamtime, but this is seen as being outside time, and from that point on, time, so to speak, is timeless. But within that, rather like Ptolemaic epicycles, are endless small cycles of generational relationships, always returning to the same place through complex rules about kinship and marriage.

Their view of the environment is the same. The environment is in a steady state. Plants and animals are always there, because, since the Dreamtime, plants and animals have always been there. The bush looks the way the bush looks and always will look. This was something of a self-fulfilling prophecy: because people believed the bush should always look the same, they made sure it did, on average, always look the same. Within the sameness, though, were cycles of succession, endlessly returning to the same point, the point of the most rich and diverse environment a particular area could support.

The acceptance (at one level) and use of archaeological evidence by Aboriginal people to achieve the goals of recognition of prior ownership of the continent, and of dispelling the white paradigm of a people who hadn't arrived long ago, and/or had done nothing with the country and retained a primitive society and culture, creates its own ambiguity. The dilemma is something like that faced by fundamentalist Christians. In rejecting evolution and an age of the earth greater than 4004 BC, you are not just saying that some aspect of geological interpretation is wrong, but that *all* science is wrong. Physics, chemistry, biology and astronomy all support geological interpretation, and to say that they are wrong is, logically, to deny the evidence of the senses that nuclear bombs, plastics, improved crops, and space shuttles all exist.

The problem is that Aboriginal people have taken over a concept that they have always been here, which was based on a particular sense of social structure, and within each particular group. It could never have been a general statement about time-depth, nor could it have been a general statement about Aboriginal people as a whole. The idea of 50 000 years, however, has been taken on by Aborigines without recognition that if you are going to accept this figure, then (a) you are accepting this as a maximum figure, and (b) you are accepting that Aboriginal people changed, the country changed, people moved, human biology changed, technology changed, culture changed over that period. It isn't good enough to simply take 50 000 as a minimum figure on the assumption that eventually archaeology will confirm that Aboriginal people have been here 'forever'. This doesn't mean that archaeology never gets things wrong. Archaeologists are, I hope, a changing people in a changing intellectual climate. But the debate has to take place within the parameters of the discipline.

Apart from the land-ownership aspect of this ideology, it has also affected the question of human remains. A logical consequence of the belief that our people have always occupied this piece of ground is that any humans found buried in this piece of ground, no matter how old, are the ancestors of the present-day owners—who else could they be? The practical outcome of this belief in recent years has been the return to the local communities of the Mungo remains, over 30 000 years old, and the Kow Swamp[7] remains, some 10 000 years old. The return of these remains has caused anguish among some archaeologists, who have argued that such remains have nothing to do with present-day people and are so old as to be within the general interest of all humanity.

In England recently, just for fun, an experiment was done to compare the DNA of the modern people living in the south-west with DNA from the bones of a Pleistocene age skeleton from the area.[8] Astonishingly, a local schoolteacher turned out to have DNA extremely close to that of the skeleton from thousands of years ago, and so was presumably, in some sense, a descendant. But it isn't surprising: if you go back far enough, then everyone is related. Even if you go back only a few hundred years, everyone turns out to be a descendant of any famous figure you care to name. If this is true of the mixture of peoples and their movements over time in Europe, how much more relevant it is in Australia. There is no doubt that the DNA that originally arrived in Australia is still the DNA being carried by all Aborigines. Nor is there any doubt that it was also the DNA carried by the Mungo people and the Kow Swamp people, and all the other Aboriginal ancestors. There is a community of relationship both across Australia and through time in which all Aboriginal people are related to each other more closely than to any part of the outside world. In the Dreamtime, that community of relationship was also established as including the animals, the plants, and features of the land.

Religious beliefs have always had a very uneasy relationship with the study of the past. In Western religious systems (that is, those arising in the Middle East, or deriving from such systems), at least until recent times, the study of the past has been at best unnecessary or at worst dangerous. Among fundamentalists of all persuasions, this is still the case. The stories of humans were set down once and for all and, because they derive from a god, are immutable. Studies of the past can and do lead to heresy, and are therefore best discouraged or totally forbidden. Less fundamentalist religious people have adopted the comforting but spurious belief that the world of science and that of religion are totally separate and each has its own truth.[9]

Aboriginal attitudes to the past are something of a combination of all these views, particularly in recent times with the addition of an enormous interest in genealogy, and a belief that science couldn't investigate the Dreamtime or the origins of human life in Australia, but that it was all very well in demonstrating a kind of deep history in a particular region. Yet generally speaking, and totally so in the past, Aboriginal belief in the truth of the Dreamtime was as absolute as that of any fundamentalists anywhere. To excavate the ground would have been clearly futile, because Aborigines knew they were an unchanging people in an unchanging land. And just as believers in the nineteenth century (and even today) saw fossils as some kind of odd geological phenomenon, or a test of faith deliberately put there by God, so too might Aborigines have viewed the exposure of megafaunal bones in a river bank, or engravings that could not have been done by any humans they knew, because there were no meanings attached to them.

In fact, and unlike the situation in the West, any accidental excavation (and deliberate excavation would have had no purpose) could only have confirmed the reality that the past consisted of just two kinds of history: that of people exactly like themselves in the recent past, and that of an earlier past so different from today that it was part of the Dreamtime. Ironically, this, in fact, is not a bad summary of the results of a century of archaeological research in Australia.

In recent times, history has become powerful and essential in Australia. The concept of the relationship between archaeological research and land rights claims is well established, though the nature of that link seems to me curious. I've never understood why 50 000 years should have been seen as providing more legitimacy to land rights claims than 20 000 or 10 000, or even for that matter 5000 years, and it has long been known that at least that order of magnitude was involved. For the descendants of the promoters of *terra nullius*, and the ancestors of the architects of the Ten Point Plan, no length of time of occupation has been seen as long enough to equate with 200 years of occupation. The proposition is a bit like the old equation that, in terms of media interest, one death in Sydney equals ten in Britain equals a hundred in France, equals 10 000 in Rwanda. It is easy to lose sight of the fact that 5000 years ago the British had a level of social organisation and technology little different to those of the Aborigines of the day. The pyramids were not yet built in Egypt.

In any case, having established that length of occupation for the continent, the idea of using excavations within a particular area as the supporting evidence for a particular claim seems ludicrous. Perhaps as archaeologists we have not got this concept across very well—occupation anywhere in the continent 50 000 years ago means occupation everywhere in the continent 50 000 years ago. Undoubtedly parts of the continent will have been uninhabited at intervals of nasty climatic conditions, but this means nothing in the overall context of continental occupation. Perhaps the debates about desert or highland or Tasmanian occupation patterns have made it look as if there were debates about whether parts of the country had never been settled, or 'only' settled at a time before the first pyramids were built. There is a similarity to the 'debate' about evolution, with the creationists seizing on a discussion about particularities of the record or processes to claim that the whole theory is in doubt. Whatever, it is time for a clear statement from the archaeological community that every inch of this continent had Aboriginal footprints on it effectively right from the moment that there were beings with human footprints.

A more explicable, though equally tenuous, link is between history and land rights. Aboriginal people in their thousands are exploring their histories, finding old photographs and documents, and building long genealogies extending well back into the nineteenth century. Aboriginal people seldom have any doubt about their ancestry or geographical origins, and requiring them to produce these few scraps seems somewhat offensive. On the plus side, the general interest in oral and documentary history by Aboriginal and non-Aboriginal people has greatly fleshed out the 'official' Australian history. Aboriginal people in history books are no longer like the Dharug in 1788, recorded only as shadowy figures in the forest by the colonists, but people (including the Dharug) with history and names and actions. This historical research was clearly not a traditional practice, and the apparently ahistorical nature of Aboriginal culture for a long time kept the two traditions, Aboriginal and non-Aboriginal, from meshing with each other. The new interest has meant that there are no longer two histories but one.

3

'A slow strangulation of the mind?':

Eating fish is wrong

Like a blow to the heart, it took a long time to take effect, but slowly but surely there was a simplification in the tool kit, a diminution in the range of foods eaten, perhaps a squeezing of intellectuality . . . Were 4000 people enough to propel forever the cultural evidence of Late Pleistocene Australia? Even if Abel Tasman had not sailed the winds of the Roaring Forties in 1642, were they in fact doomed—doomed to a slow strangulation of the mind?

—RHYS JONES, 1977[1]

The idea that Australia is a continent full of doomed people, animals and plants is a theme that recurs over a very long time. It may be that the idea appeals, and certainly this is made explicit by some people, because it assuages guilt and provides an alibi. Landing on an unspoilt continent and proceeding to wreck millions of years of evolution, a fragile environment, and 50 000 years of culture is a pretty heavy load of guilt for white Australians to carry. Much better to think it was only coincidence that we arrived when we did. People, animals and plants were doomed anyway— just not up to the job; and we didn't cause the problems, we simply observed them. Furthermore, there is no point in being sentimental about this, and no point in investing time and money trying to save doomed things. Much cleaner and quicker to let them go, and kinder too, really. So Tasmanian Tigers are gone? Nothing we could do about that; they had been declining since the Tertiary period millions of years ago. Other small mammals and birds are on the verge of extinction—don't waste your money. Tasmanians extinct? Well, they had it coming, only had a few years left anyway. Could be thought of like the euthanasia of species and races.

But people, and plant and animal species, don't just become extinct as a result of some genetic predisposition from evolutionary history, or some social or cultural inadequacy. They become extinct because somebody or something caused the extinction.

Tasmania is a good test case of this kind of hypothesis, and it is also something of a microcosm of Australian archaeology, and has played a significant role in discussions on Aboriginal origins. The Tasmanians were the people Pulleine was referring to in his 'unchanging people' comment. But they had certainly changed in at least one respect, and this has become one of the top ten Australian archaeological mysteries. Astonishingly for an Aboriginal group living on or near a long coastline, Tasmanians didn't eat fish. Rhys Jones discovered that their ancestors did eat fish but had stopped doing so some 3000 years ago. The evidence was from Rocky Cape, but whatever the reason, all over Tasmania people had apparently stopped eating fish around the same time. This change didn't solve the question; if anything it made it even more puzzling.

Rhys Jones developed a theory that he believed fitted the facts.[2] He suggested that the culture of the Tasmanian Aborigines had become 'depauperate' because of the long isolation of Tasmania from the Australian mainland. He believed that some tools and weapons had gone out of use. There was archaeological evidence that bone points had been in use at one time but not more recently. Items such as boomerangs and spear-throwers, probably in use before Tasmania was separated from the mainland, were not in use in Tasmania in modern times, and so may have been lost along the way.

The problem with these examples was that there was no archaeological evidence for their ever having been in use in Tasmania, though ancient boomerangs being found in southern South Australia made it seem likely they had once been in Tasmania. While the total tool kit known ethnographically in Tasmania does appear to be relatively small (though the difference is not as great as is generally suggested), there is no evidence apart from the bone points that this is the result of the loss of items from the mainland tool kit.

But fish (and bone points) had undoubtedly been lost, and the question is why. Jones feels 'that the only possible explanation is that the Tasmanians made an intellectual decision which had the result of constricting their ecological universe'. That intellectual decision, he thought, was in the 'area of prohibition and sanction'. So what? Well, not using an important food and not using bone points were evidence, he believed, that Tasmanian culture was deteriorating in the absence of outside stimulation, evidence of 'slow strangulation of the mind'. In other words, some kind of religious movement had come over the Tasmanians and, like all religious movements, it had led to irrational behaviour. Not only did the prohibition bring no benefit (and there are religious prohibitions of certain foods all over the world, that are of benefit to the people concerned in many different ways, such as preventing food poisoning in hot climates), but it actually did them harm by reducing their food supply. Not enough intelligence, then, to understand that not eating fish was a stupid move.

But the past is never amenable to simple explanations, and Jones himself quickly spotted the first flaw. If the tool kit was simplifying and food items were dropping out of the diet, the obvious consequence should be that population density would fall too, as the ability to extract subsistence from a given area of land was reduced. In the end, presumably, though Jones doesn't develop the proposition, population levels would fall so far that the Tasmanians would become extinct. But the facts didn't fit. Tasmanian population density was not just equivalent to the mainland Australian scale, it was at the high end.

So what was going on? Jones went back to the drawing board and came up with a different answer (and one that would repay a great deal more analysis than it has received).[3] Something had to give in Tasmania, and if it wasn't population, it must have been the non-utilitarian aspects of society. More generally, Jones believed that the apparent increase in complexity of the mainland tools had given Australian Aborigines more spare time after earning a living, and this spare time could be used for cultural and artistic pursuits. It is hard to know what the direction of the argument is here. It could also be turned around to suggest that Tasmanians had been forced to work harder and harder, spending more and more time on economic life and less and less time on cultural life. In fact, although the argument has not been made, it could be suggested that the reduction in cultural life would add to and accelerate the 'slow strangulation of the mind'.

And the evidence? Well, Jones contended that 'large scale religious events, such as are described for mainland society, were not part of Tasmanian cultural behaviour'. But the whole argument began to get messy when Harry Lourandos suggested that in fact Tasmanian population densities *were* lower than those of the comparable Victorian environment.[4] So there may be no need for the proposition that Tasmanian Aborigines were drudges without labour-saving devices, while their cousins on the mainland, with all modern conveniences, could sing and dance all day. But we will come back later to the proposition of a difference in cultural life between Tasmania and the mainland. First we need to look at the general proposition of a difference in economic life. Did the Tasmanians have a simpler tool kit and, if so, why? They undoubtedly stopped eating fish (this is the only fact we are dealing with in the whole debate), but why, and so what?

If a picture is worth a thousand words, the pair of pictures that Rhys Jones presented to show the richness of the mainland 'tool kit', and the poverty of the Tasmanian one, were worth tens of thousands of words.[5] On one side just seven items, on the other side thirty-one. But what do these items tell us about the economic life, let alone the intellectual capacity, of the Tasmanians?

There is little doubt that if you ranged the entire technology of a group from south-east mainland Australia against that available to a Tasmanian group, the latter would probably seem relatively limited in total, and at least some individual items might be relatively more simple. But keep in mind that there isn't much in it. A Kurnai man transported to Tasmania would have had little trouble earning a living with the tools he was given, and a Pyemmairrener woman dropped on the shores of Gippsland would have quickly made herself at home. The former might have missed his boomerang and made derogatory remarks about the workmanship of his skin cloak; the latter might have grizzled about the clutter, and about a lot of fuss to do a simple job. Neither would have thought of themselves as being transported into an alien civilisation.

But granted a difference, why is it so? The problem in answering the question is that it is the wrong question, and the act of posing it directs the answer into the wrong channels. People tend to think that the art of science is to find answers, whereas finding the right questions can be of equal or greater importance.

Aboriginal Australia can be divided into eighteen distinct regions.[6] The peoples of each region differ from peoples in other regions in a variety of ways—artistically, religiously, technologically, socially, linguistically, musically and so on. The particular set of cultural attributes in each region is related to the climate and topography, and ecology, and to historical factors. It is no surprise that the people of the desert differ from the people of the rainforest, or the riverine plains, or the Kimberleys, or the Torres Strait. It would be surprising to find that the people of Tasmania were not culturally unique also. In some regions, the aspects of the environment that relate to the cultural diversity are obvious, in others less so. What was it about Tasmania that resulted in the Tasmanian culture?

For a start, Tasmanians didn't eat fish, so a whole range of nets and spears and other devices weren't needed. In areas of thick vegetation, spear-throwers and boomerangs would have been of little value. There was little opportunity for processing cereal and other grains. Conversely, in a place where you only had to reach into a burrow to pull out a mutton bird chick, where the brush could be so thick that wallabies were restricted to such narrow trails that a sharpened spike stuck in the ground would impale them, and where you only had to crawl into a seal colony and pretend to be a seal for a while before bashing them over the head with a club, there was not much need for sophisticated food-procuring technology. At least some of the technology of mainland Australia was the result of the need to kill animals whose presence was difficult to predict, at relatively long distances, in conditions of open plains or woodlands where animals could see a long way and were free to follow any escape route.

In a rich environment a simple technology is sufficient. The absences from the 'simplest tool kit in the world' were a result of refining that tool kit to meet the needs of the Tasmanian environment. The Tasmanians may have been the first to practise Thoreau's dictum, 'our life is frittered away by detail . . . simplify, simplify'. Why would you put effort into developing equipment that was not needed? And there is a curious twist to the proposition that Tasmanians were abandoning items of material culture because of intellectual decay in isolation. One of the great mysteries of Australian prehistory is that backed blades were developed several thousand years ago (though not in isolated Tasmania) and could have been seen as another example of the dynamic nature of Australian technology in contrast to Tasmanian. Here was a simple flake technology giving way to a new and more complex form, while Tasmania stayed with the old simple flakes. Unfortunately, around 1000 years ago, backed blade manufacture was abandoned on the mainland and people went back to the simple flakes. Whatever the reason, this is presumably not an example of strangulation of the mind over the whole Australian continent. Why, then, should the loss of bone points be seen in this way in Tasmania?

And so to fish. Prohibitions on foods are known all over the world and at all different times in history, so why does the Tasmanian prohibition of fish seem odd to Rhys Jones? Again, there is no simple answer to this question, and this is partly because Jones changed his mind, the change indicating one of the crucial aspects of the affair. In 1971 he suggested that dropping fish from the Tasmanian diet was no big deal. There was 'no shortage of protein . . . with abundant seals, shellfish, birds and land mammals'. Carbohydrate was the big problem. So 'the prohibition of the eating of fish could be tolerated by the

Tasmanians because although it may have been inconvenient, it did not vitally affect their livelihood'.[7] The point here was that not eating fish was a kind of quirk of fashion that you could drop on a whim, and was therefore evidence of the intellectual strangulation that was going on in isolation. Presumably, if fish were vital, there would be economic pressure to keep them, and no matter how restricted your intellectual universe you would go on eating them.

But if they were so unimportant, then the quirk wouldn't tell you much at all. Elsewhere in the world there are prohibitions on foods for religious reasons, and within Australia there are endless examples of food restrictions for certain groups. So by 1976 Jones had changed the argument. Now fish were seen as being potentially crucial during the long hard days of a Tasmanian winter, and their absence from the diet would have caused 'significant deprivation'.[8] Now we are in a new ball game. Now we have evidence of a real loss of intellectuality—Tasmanians were obviously a few fish short of a full net. Religion was one thing, but starving yourself was evidence that things were going seriously downhill.

Now the argument is complete. While all over the world different groups choose to eat or not eat various food items, Jones thinks that the absence of fish-eating in Tasmania is of much greater significance than any other dietary oddities. For it to be the result of isolation it has to be unique, and Jones presents a number of reasons that make it seem unique to him.

It was, apparently, a prohibition that applied to all members of the society, not just certain subgroups (for example, pregnant women, or old people, or children, or particular kinship groups), and, indeed, apparently applied to all the groups in Tasmania. This was significant because it meant that the reason was presumably something associated with the island of Tasmania as a whole, rather than some cultural or environmental quirk of part of the island. It also apparently applied to a whole class of animals, rather than, say, a particular species of kangaroo with totemic significance to some people. The Tasmanians were mainly a coastal people, and the animals prohibited were fish, of enormous value to coastal people all over the world in all periods.

But such economic value could be overcome by some overwhelming environmental consideration. What if the fish had become extinct, for example, or become toxic? What if massive storms had damaged the reefs? What if changing currents had changed temperatures and therefore the marine ecosystem? There is no evidence for any of this, but the only evidence against is the fact that, as Jones pointed out in 1978, 'the same wrasses and leatherheads swim around the rocks and shallows of Rocky Cape and Little Swanport respectively as was obviously the case eight and five thousand years ago'.[9]

So if we can eliminate the environment (but note that the fact that the fish are there now doesn't contradict some of the imaginary scenarios outlined above), then it leaves a stupid decision as the cause—something that seemed like a good idea at the time. But it is at this point that doubts start to creep in. Okay, so some religious movement sweeps through one of the Tasmanian Aboriginal groups 3000 years ago, and the elders decide no more fish. But did none of the other groups say, 'hang on a minute, you guys, you can give up fish if you like, but don't come crawling to us when you are hungry next winter'? I am reminded of a comedy sketch I heard once about Walter Raleigh bringing tobacco back to England for the first time. 'So what do you call this stuff, Walt? Tobacco? And what do you do, you roll it up, right, and then you stick it in your mouth? Yeah Walt, and then you do what? You set fire to it? Riiiight! That's really going to catch on, Walt!!'

You would think that at some point in the subsequent 3000 years someone would have spotted that the Emperor had no clothes. So what was going on? Well, I want to explore in more detail two of the aspects touched on above: was the absence of fish of any importance in the Tasmanian diet, and was there really no kind of environmental factor involved?

Ron Vanderwal excavated a very interesting set of sites around Louisa Bay and Maatsuyker Island in south-west Tasmania, enough of a range to give an idea of seasonal patterns in the economy.[10] In brief, the yearly round involved travelling to an offshore island to get seals and seabirds in summer, and hunting land mammals, and the occasional beached seal on the mainland, in winter. While Rhys Jones saw the Tasmanians as having 'summer plenty, winter stress', it may be that this difference was a fairly recent development after seal numbers declined in Tasmania. Before that the difference was that in summer you went to the seals, in winter they came to you; in summer you ate seabirds, in winter wallabies. But there were certainly more seals available in summer, and food would have been relatively more plentiful then.

But fish were obviously not much of a loss. In the summer, with large numbers of seals and seabirds available, fishing was unnecessary. During the cold winters, seal flesh would have provided a much richer energy source than fish, and there was the added advantage that seals came out of the water, and so were hunted on land. Entering the water in winter in south-west Tasmania would have been both hazardous (because of the risk of storms) and inefficient (because of heat lost from the body, which would be hard to replace with fish meat). As long as the seals kept coming during winter, and land mammals were available as a staple, fishing would have provided many disadvantages and dangers and few advantages. I am not suggesting that there was a conscious adding of pros and cons and a 'decision' taken. As we shall see later in relation to agriculture, I don't think societies take over economies, I think economies take over societies, a process which continues today! My guess is that the fish prohibition was transmitted from the west with religious overtones, and became adopted as a religious belief without the economic rationale that had started the process (climate and sea conditions are milder than in the west, but neither fish nor seals seem to have been eaten).

We know that ceremonies developed in one part of Tasmania could be rapidly adopted in other parts (just as could happen in mainland Australia), a famous example being the 'horse dance'. In the south-west, a number of incidents were recorded by George Augustus Robinson, in which there were a whole series of prohibitions on cooking 'mutton fish', burning human hair, burning human bones, making a basket for human bones, roasting goanna, all of which were said to make the rain come. You might well not want any more rain in south-west Tasmania than you had to have, but the strength of the reactions to these actions suggests that 'rain' was a shorthand (or the best Robinson could understand) for storms.[11]

This is confirmed by three other records from Robinson: 'The Brune natives said that the Brune devil is thunder and lightning'; 'the song was the devil's song and their homage is the homage paid to the fire spirits'; 'before [the women] plunge into the water, they stand on the rocks in rather an obscene position and chant a song.' Such rituals and beliefs are involved with the powerful forces that affected life in Tasmania—storms (which on land made hunting difficult, life uncomfortable and even dangerous, and at sea could kill people) and fire (which was a positive force bringing warmth to counteract the storm). I think the prohibition against fish-eating, developed in western Tasmania because of hardship and danger, became part of the storm–sea–fire religious complex, and spread in this form to the rest of Tasmania.

In the east, for example, there was a myth about stingrays. Two women

> were swimming in the water, they were diving for crayfish. A sting-ray lay concealed in the hollow of a rock . . . he spied the women, he saw them dive: he pierced them with his spear—he killed them, he carried them away . . . The sting-ray returned, he came close in shore, he lay still in water, near the sandy beach; with him were the women, they were fast on his spear . . . The two blackmen fought the sting-ray; they slew him with their spears . . . the two blackmen made a fire . . . on either side they laid a woman—the fire was between: the women were dead.[12]

And not just a myth, but a great associated ceremony where 300 people were seen surrounding stingrays and spearing them, but not eating them.[13]

Once such religious beliefs had become established throughout Tasmania, they would effectively prevent the exploitation of fish even where the economic imperatives for not eating fish were much weaker or absent. Eventually, scale fish came to be seen as 'not food', and this was the end of it. In parts of Africa, swallows were once seen as food (and eaten in huge numbers) during their annual migration from Europe. We don't see them as food when they fly around our houses. Nor would the English now see larks and other small birds as food, though they were once a great delicacy, and still are in parts of Europe. In some areas grasshoppers are eaten, but again, in the West, we see a hopping insect in the grass as belonging to the 'non-food' category. The Tasmanian, reaching for an oyster, and seeing a fish swimming past, would have been no more confused about which was a food item than we are in reaching for biscuits and seeing a mouse, or reaching for a cabbage and seeing a slug.

Rhys Jones thought that the complex tool kit on the mainland gave you time for a much richer ceremonial life than was possible in Tasmania. But I think the reverse is true, and that in Tasmania the tool kit was simplified, and the combination of this and a rich predictable environment gave you time for ceremonial life. So, did they use it?

There wasn't much chance of Robinson recording ceremonial life: he was travelling, not living in a settled community, travelling with strangers, and he had no idea, as a highly religious man, of any ceremonies except those of the Anglican Church. He was also watching a society under enormous pressure. It is surprising not that Robinson recorded so little ceremonial activity, but that he recorded so much, and what he recorded is clearly just the tip of the iceberg of a rich and complex cultural life in Aboriginal Tasmania.

There were dances and storytelling that continued for days, a whole range of rituals associated with conception, the arrival of strangers, hunting, courtship, the arrival of whitefellas, battles. Many of them involved large numbers (two records specifically of 300 people). There were rock carvings and ceremonial sculptures. The rapid adoption of new ceremonies across the island, including the fish prohibition, is a sign of a great vitality in social and cultural life.

Rhys Jones saw the difference between Tasmania and mainland Australia as being much greater than any of the differences that occur between different areas within mainland Australia. He suggested that this uniqueness is expressed in both economic and cultural terms, while I think the cultural life of both was similar, and that the economic difference was a response to the environment of Tasmania.

Tasmanian culture and economy was certainly different to that of mainland Australia, but to claim that these differences confer uniqueness on the Tasmanians is to make a value judgement as to the relative importance of various cultural and economic traits. To take some examples at random. Why are people who do not eat fish more unique than people who collect Bogong moths, or hunt dugongs, or grind seeds? Why is not using boomerangs at all a more significant absence than the lack of hunting boomerangs in northern Australia? Is the difference in clothing between Tasmanian and Victorian Aborigines any greater than between south-western Australia and Victoria? If it was true that there were fewer large-scale rituals, why would this be seen as more significant than the lack of the circumcision ceremony in parts of the mainland?

My map of Aboriginal Australia divides the mainland plus Torres Strait into seventeen regions. The eighteenth region, Tasmania, is the only one whose boundaries coincide with a clearly recognisable geographic barrier. Bass Strait gives us our ability to recognise 'uniqueness'; it doesn't cause uniqueness by forming a barrier that leads to some kind of cultural decay.

If the sea had never risen to cover Bass Strait, there would still have been a Tasmanian region with pretty much the characteristics (though the presence of the dingo might have enabled better exploitation of the rainforest) that George Robinson saw in 1830 and that Rhys Jones attributed to the effects of isolation. Let us imagine Cape York separated by sea from the rest of Australia. Would the absence of skin cloaks and hunting boomerangs in the two Cape York regions then have been seen as examples of a decline in intellectuality caused by isolation? I don't think so.

And the moral of this story? Well, three really (a fourth, 'beware ethnography', is too obvious to need expansion). First, people are pretty much just people, and a human life is pretty much human life. If we paint Tasmanians or Tierra del Fuegians as some kind of second-division humans, with reduced intellectuality and an impoverished life with little culture, this is in effect the same kind of process as that which characterises Bosnians or black slaves or Jews or Gypsies as being sub-human. We need to recognise that all humans have an equally full life, though they may have different elements in different proportions, and they will certainly do things in different ways.

Second, never discount the environment as an agent in human affairs. There may be aspects of human life that seem to be purely related to the realms of the mind or the soul, but just when you think you have found an area where human activity is above Nature, along comes Nature and pokes you in the eye.

Finally, indigenous people are smart about their environment. It would have been easy to use fish and the Tasmanians as an example of indigenous people being no better than the rest of us when it comes to recognising where their best interests lie in relation to the environment. But it isn't. It's another example of balancing competing needs and forces. It is hard to find examples of indigenous people not doing that, and even if this ability is not 'instinctive', it is nevertheless clearly strongly linked to the human condition.

And there is another kind of moral to this story. No one, human, plant or animal, is 'doomed to extinction'. It is an unfortunate misconception, and one that is increasingly used, again to provide an alibi for bad behaviour. It is not that long ago we were soothing the pillow of the Aborigines as a dying race, inevitably doomed—as were indigenous people everywhere— by the magical properties of Western civilisation as it came into contact. In the environment, the same argument has begun to be used about particular species which are representatives of taxo-nomic groups that have become greatly reduced in numbers. We shouldn't, the argument goes, worry about trying to save them— wasted effort, let them go (nicely of course). But the fact that related species have become extinct is no predictor of whether a particular species will go. There is no more a kind of taxonomic vigour than there is a racial vigour which ensures the triumph of some races and not others. The Tasmanian Tiger may have been the last of its group, but it would be surviving now if it hadn't been hunted to extinction by men with guns. Platypus and echidna may speciate and increase their group size again, but whether they do or not has no implications for their survival, which depends on the maintenance of suitable environments for them.

The lack of fish-eating in Tasmania wasn't the result of the intellectual impoverishment of a small group living in isolation, but of economic and environmental factors that became codified (as they have in many other parts of the world) as religious prohibitions. The original rationale becomes forgotten in time, but the reasons remain good ones.

Tasmania wasn't different to the rest of Australia, but in one particular respect Australia has been seen as different to the rest of the world. In a country that is now a major supplier of food to the rest of the world, there was apparently no indigenous agricultural development in Australia. Just as with Tasmania, on a larger scale, one reason that has been suggested for this amazing absence is the isolation of the whole continent from the rest of the world. Isolation doesn't really explain anything in Tasmania. Does it for Australia?

4

'A people so inclined':

To farm or not to farm

We are to consider that we see this country in the pure state of nature, the Industry of man has had nothing to do with any part of it . . . In this extensive country it can never be doubted but what most sorts of grain, fruits, roots etc of every kind would flourish here were they once brought hither, planted and cultivated by the hand of Industry.

—JAMES COOK, 1770[1]

Native birds such as geese, pelicans or scrub turkeys, might have been domesticated by a people so inclined.

—JOSEPHINE FLOOD[2]

It is sometimes thought that Aborigines are unique because they were hunter-gatherers. This is not true; even in Europe there were probably still pockets of people supplementing their food by hunting and gathering not that long ago, and all the other continents had hunter-gatherers 200 years ago. No, the uniqueness was that this was the only continent without indigenous agriculture. The reason for this has been hotly debated, and both the truth of the observation and its implications have political ramifications. In this chapter I explore the nature of the Aboriginal economy and the question of agriculture.

In all parts of Australia, all Aboriginal people were hunter-gatherers before 1788 (and for a long time afterwards in most parts of the country). Their food came partly from hunting, using often complex instruments, such as a weapon or a trap to catch mobile large animals that could not be caught with the bare hands. The bulk of the food came from gathering, collecting relatively small plant products and immobile animals, either by hand or with generally simple instruments. There was something of a gender division between the two activities, men seeing themselves as being hunters, of course. But women would frequently collect fairly small animals such as rodents, bandicoots and reptiles, either by hand or with a simple digging stick. Men would gather (and often eat on the spot) such foods as berries and shellfish while hunting. While fish could be hunted with spears, they could also be gathered by both men and women using stone tidal traps. In some areas quite large animals (a notable example being seals in Tasmania) could be hunted by both sexes.

Hunting was not only enjoyable and exciting and prestigious, but occasionally, with luck, could bring a welcome influx of fresh meat into a camp. But you couldn't rely on it, and the more mundane items, such as seeds, roots, berries, eggs and shellfish, which could be obtained every day, all year round (if you timed your movements right), were needed to keep the stomachs full and the children growing. It is hard to be precise about the relative amounts of plant and animal food in the diet, because it would vary so much between coast and inland, arid and wet lands, high and low lands, and seasonally. Possibly, plants provided the bulk of the diet in most places. There is less debate about the relative contributions of men and women to keeping food on the table—women by a country mile. But, now and again, when the roulette wheel stopped in the right place, in would come the men with enough food for a feast.

The diet everywhere in Australia was well balanced, healthy, and normally of sufficient abundance, though not enough to make people fat. Although the lack of body fat was at least partly due to the expenditure of energy in obtaining food, in most parts of

Australia the diet was low in fat (exceptions being areas like Tasmania and south-eastern Australia, with a seasonal reliance on seals and mutton birds), although fairly high in carbohydrate. The diet pyramid consisted of a top 10 per cent of large animals like kangaroo and seal, and a middle 30 per cent of medium-sized animals like wallabies, possums, birds and large fish and reptiles. The bottom of the pyramid, perhaps 60 per cent, consisted of small mammals, birds, reptiles and fish, insects, shellfish and eggs, and seeds, roots, berries, nuts and other plant foods.

The diet was considerably better than anything the first Europeans in Australia had (or had come from), and indeed it was healthier than European diets throughout at least the nineteenth century. But there were lean seasons, in winter in the south, in summer in the centre, in the wet season in the north. Not only could resources be scarce at such times, but environmental conditions could make obtaining them difficult for people. In Tasmania, winter storms and cold made getting marine resources difficult and dangerous; in the north, storms were also a factor in both restricting movement and actually destroying resources like shellfish beds, and lowland flooding could make travel difficult. In the desert, temporary water sources dried up in summer, so people had to live near permanent water and were consequently restricted in how far they could go to get food. Food supplies available within a day's travel would rapidly become used up.

In lean times, people would be forced to make do with foods they would normally have ignored. In good times, highly prized and tasty foods would be sought. Whatever the season there was an attempt to keep the diet as balanced as possible. But people did not simply have an open-ended choice of every potential food item. Social groups such as children, the elderly, and pregnant or menstruating women were required to eat some foods and were denied others. Food obtained from hunting was distributed according to strict rules relating to kinship. Some species of animals were the exclusive hunting preserve of some social groups, and other species could not be eaten at all by other social groups. Most areas had some food which became seasonally superabundant, famous examples including bogong moths and bunya nuts. Such foods could support large gatherings of people for major ceremonial events.

We tend to think of hunter-gatherer diets as being much more varied than Western diets, and they are. The difference is not so much in the *classes* of animal eaten—a Western diet after all may include oysters, snails, crabs, lobsters, honey, fish, crocodile, poultry, sheep, cattle and deer—but rather in the number of species in each class. Hunter-gatherers will exploit many different birds, many different mammals, many different shellfish, and so on. With one exception, Western diets tend to include only one or two species in each class that are considered to be food. The exception is fish. Western diets include a much greater range of deep-sea fish than any hunter-gatherer could have had access to. On the other hand, the hunter-gatherer would make use of more in-shore and reef marine species, and certainly more freshwater species than can be seen in fish shops.

Hunter-gatherers tend to waste very little that is edible, and consequently get a good range of nutrients from their food (for example, eating small animals whole, and consuming all edible parts of large animals). They also avoid losing precious nutrients during food preparation, with food being lightly baked or roasted, not boiled. Since food was rarely stored in Australia, most foods being eaten fresh, there was little loss of nutrients from storage or preservation techniques. Similarly, all the animals and plants consumed were non-domesticated, an advantage in lower fat content, but perhaps with lower nutritional value in some cases. In almost every respect, the diet of hunter-gatherers in Australia was very healthy for people. Was it also healthy for the environment?

A hunter trying to feed wife and kids and the rest of the extended family can't worry too much about the resources being harvested, nor be too sentimental about them. Birds are easier to catch, and their eggs and young obtained, when they are breeding in colonies. So are seals and their young. Spawning fish can be easy targets in shallow water. While large animals like kangaroos would generally be obtained only one at a time, smaller wallabies could be collected in traps in considerable numbers after drives. Fish could also be collected in large numbers in traps, and the traps themselves could affect water movement and siltation. Travel to offshore islands to get marine food could inadvertently introduce damaging animals such as the dingo and rats. But removal of the weak and old and excess young from a population might help to strengthen that population.

Sedentary animals were even easier to catch, of course, and nearly all the shellfish could be removed from a particular area. While whole plants were rarely removed, much of the plant material used was reproductive material—seeds, tubers, bulbs, fruits—and this could affect numbers the following year. This effect could be compounded because people needed to live near water sources. The surrounding area could be denuded, including the removal of both dry and green timber for fires. On the other hand, digging up root crops cultivates the ground and breaks pieces off, which multiplies the number of plants, and carrying plant reproductive material back to camp will help to spread species.

As well as there being something of a balance between positive and negative effects, a number of factors combined to minimise the impact of Aborigines on the Australian environment. They lived most of the year in small groups in large areas, and hunting and gathering were done only for the immediate needs of the group. With no storage or trade in food there was no incentive to kill or gather more than was needed for the meals of the day. Even if there had been the incentive, generally speaking the technology, whether spearing kangaroos, smoking out possums or digging tubers, was not of a kind to allow mass harvest. Even if there had been the means, the lack of land transport would have reduced the ability to move quantities of food large distances—killing large quantities of kangaroos in a single hunt would make no sense.

Finally, and it is a crucially important factor, Aboriginal mobility protected the environment. If you are not in permanent fixed communities, then when a resource has been exploited for a while its numbers will be reduced to uneconomic levels, to the point where more energy is expended in obtaining it than is gained from it. Resource levels also change in response to seasonal conditions—no single resource is harvested all year round. For both reasons, hunter-gatherers move periodically and systematically, and this leaves each area to regenerate for periods of a year or more before it is exploited again. You don't come back until you know there will again be sufficient resources to maintain the group for a while. While abundances in any one area will fluctuate—perhaps dramatically—over time, the pattern of movement in all the country belonging to a group is aimed at maintaining overall resource levels at a constant average. It is a pattern of behaviour familiar to us from the world of agriculture, where crops and flocks are rotated through different paddocks, and areas are left fallow in rotation to recover after cropping. So what's the difference—why weren't Aborigines farming?

Like Australia, Europe was separated from the rest of the world for most of its history, and developed its economic systems in isolation. All the economic systems of Europe were based on domesticated plants and animals, but varied enormously, from the intensive agricultural practices of Western Europe to the very non-intensive sheep-herding of the alpine regions, and the reindeer-herding of the north. Different plants and animals, and combinations of them, formed the basis of farming in different regions depending on climate, soils, technology and cultural traditions.

As Europeans broke out from their isolation, crossing deserts and oceans to visit other continents—Asia, Africa, and the Americas—they found the same pattern. The actual species involved were often strange, but everywhere you looked, people had domesticated something and were deriving at least part of their economy from it. Unlike Europe, however, all the other continents had some peoples, often in extreme environmental conditions such as deserts, jungles or ice caps, who gained all their sustenance by a combination of hunting and gathering. Such people were seen as having retained an early form of economy, either because of the environment or because they had not yet developed agriculture.

As in so much else, Australia was different. It was different not because *some* Aborigines didn't practise agriculture—it would have been obvious to William Dampier in 1688 that not much agriculture was possible in the north-west, for example—but because *none* did. This was the only continent in which there was no indigenous agriculture, and this fact is both an irritating anthropological mystery and a matter of great significance to the legal status of Australia and its indigenous people. The political imbroglio, domestic and international, about native title, Mabo, Wik and the Ten Point Plan, is the natural consequence, 228 years on, of James Cook failing to see kangaroo-herding or the harvesting of fields of native millet, or fences and walls. Indeed, as well as legal consequences, the lack of indigenous domesticated plants and animals in Australia has since 1788 coloured the perceptions of Aboriginal people held by the non-indigenous colonisers, right up to the pathetic nastiness symbolised by the views of Pauline Hanson.

There were two things about the total absence of agriculture that were surprising. One was that no indigenous agriculture, making use of Australian plants and animals, had been developed. The second was that there had been no import of any of the agricultural or horticultural systems of Southeast Asia and the Pacific. Horticulture had got as far as the Torres Strait but no further. Although these two things are usually treated as if they are part of the same question, and the same solution, they are really quite separate.

From a European perspective, agriculture looks like the natural order of things, the natural endpoint that a civilised people reach after a long period of development. It was a view confirmed when other agricultural systems were found in Asia and the Americas, based on different, indigenous species of plant and animal, but based on the same idea and the same structures and processes. Societies evolve, and eventually they evolve into agricultural societies, and then comes civilisation.

Australia's lack of agriculture was a puzzle, and there were three possible solutions. Perhaps this evolutionary model was wrong, and agriculture wasn't an inevitable development from hunting and gathering. Or Australia was simply the last continent to go through the sequence, a sequence nipped in the bud by colonisation. Or Australia was different—either the plants and animals were too primitive or too odd to be domesticated, or Aborigines were too stupid and primitive to domesticate them, or both.

The third solution was generally the one put forward in the past (and even today by some people), with varying degrees of offensiveness or academic rationalisation. It has been popular because it rationalises and justifies the legality of British colonisation and Aboriginal dispossession. The second solution has become popular among academics in recent times, though it begs the question of why development was so slow in Australia, particularly as the immense length of time of occupation has been gradually revealed. I believe the first solution is correct—agriculture isn't an inevitable development, and the pressures against it are greater than the pressures for it. But let us have a look at other attempts at a solution, including a variant on the first: people choose not to develop agriculture if they are having a good time being hunter-gatherers.

The problem with other discussions on the question is that they combine lack of indigenous agriculture with lack of introduced agriculture, as if these are just different aspects of the same question. This combination distorts the question and confuses attempts to answer it. Among the many reasons that can be put forward as to why agriculture did not spread from New Guinea to Australia (overland or by sea) is that Aboriginal people in the north made a conscious decision, on various grounds, not to change their economic system. Whether this proposition is right or wrong (and it is probably wrong), the fact that it is put forward leads to the idea that the lack of indigenous agriculture was also a conscious decision.

Generally, people reject the idea that Australia's soils, climate or animals, especially in northern Australia, were unsuitable for agriculture by Aborigines. Obviously, since agriculture is practised there now, this can't be the reason, though that begs the question of whether early, primitive agriculture could have become established. Then we turn to either cultural or economic factors. Were the Aborigines just extremely religiously conservative? No, says Peter White, after all they did change in all sorts of other ways.[3] So the answer is economic. After all, hunter-gatherers are, famously, the original affluent society, and agriculture was hard work. In this view, Aborigines saw what their northern neighbours were doing and said 'him silly bugger, eh?' Why would you slave at farming or horticulture when you didn't have to? Thanks, but no thanks, you stick to your gardens and we'll stick to hunting, and we know who is going to have the most leisure.

Josephine Flood heads into southern Australia. Okay, she is saying, given that Aborigines in the north had turned their backs on the benefits of agriculture, why had the people in southern Australia done the same? Not because they had seen other people doing it and rejected it, says Flood, but because they had weighed up the pros and cons of such a (hypothetical) system—a system that would have gone well beyond some minor activities that supplemented standard hunting and gathering—and said no thanks. Well, knocking back agriculture in the north was one thing (a lot of southern politicians and economists feel the same way), but what was going on in the south of the country, land that was to become breadbasket to the world?[4]

Flood has several goes at trying to explain the inexplicable. First she tries the 'Australian plants and animals aren't up to scratch' line—'the question of the possible domestication of Australian mammals is easily answered: there were no native marsupials suitable for domestication.' But she offers no justification for this, and it is simply untrue. Some animals would be a challenge, but even Flood finds it difficult to rationalise about some native birds and many plants ('geese, pelicans or scrub turkeys, might have been domesticated by a people so inclined'). There is no evidence that wombats and wallabies are intrinsically more difficult to domesticate than, say, goats, llama and cattle or, for that matter, wild sheep, though the differences in the reproductive system may have caused problems or required different strategies. Several of the small wallabies are used as laboratory animals for research and live and breed well in enclosures. Given the Aboriginal hunting practice of driving mobs of wallabies into enclosures in order to kill them, the step to permanently enclosing and maintaining such animals was, on the face of it, not a big one. Emus and crocodiles are now being commercially farmed (including by

Aboriginal groups) and seem to be readily domesticated. Many native plants are now being commercially exploited also.

There are several other environmental features of Australia that Flood thinks may have prevented the development of agriculture. First comes the question of food storage. Everyone knows that food storage was a problem for Aborigines and indeed for the colonists until the invention of the Coolgardie Safe and eventually the ice chest. Essentially, what Flood suggests is that if you can't store food, there is no point in producing it, the argument made above in relation to excess hunting. But many seeds and tubers and fruit *were* stored in Aboriginal Australia. The only thing you couldn't store was meat, and Flood seems to think that storing meat was a prerequisite for farming. Now certainly in parts of Europe animals were killed and their carcasses preserved, but this was because in the extremes of climate it wasn't possible to feed numbers of animals through winter. This wasn't a problem in most parts of Australia. Even if meat storage was a problem, it was a problem shared with the Middle East and the Americas.

In what is by far the most comprehensive and subtle attempt to address the 'no indigenous Australian agriculture' problem attempted up to now, Flood puts forward two other propositions that are variants on the 'affluent society' argument. While suggesting that for people in productive areas 'no doubt the labour involved in tilling and planting outweighed the possible advantages', she realises this won't wash for the arid areas—not much 'affluence' for hunter-gatherers in deserts. And furthermore, it is precisely in the arid areas of the Middle East that agriculture did develop. So why not in the Australian deserts?

Well, says Flood, in Australian deserts, 'when food was difficult to obtain, the food quest simply required more time and effort rather than new strategies'. And why was this? Apparently 'there was no pressure on the amount of land available, in strong contrast to the Middle East and the narrow neck of Mexico'. I guess what she means by this is that Australia, with its wide open spaces, allows people to just go further and further in search of food when times are tough. In small areas (the Middle East!) you bump into your neighbours, so instead of just wandering further you have to come up with a new approach. Eureka! Agriculture! But of course you also bump into neighbours in Australia, and there aren't (in spite of the populate or perish brigade), or weren't, great unused areas you could go to. Furthermore, when times are tough, water is scarce, and your ability to wander far afield is very much curtailed, as you camp by the rare scattered permanent water sources. In fact, this is an 'oasis' situation much like what has been proposed for agricultural development in the Middle East. But there is a germ of an idea here, and we will come back to it.

Jo Flood's other interesting idea is to do with environmental predictability. Investing in agriculture, she says, requires you to know what the future holds. She seems to think that the Australian environment is the most unpredictable of the inhabited continents, and that this in turn promotes an absence of risk-taking. If you get a drought following a flood and followed in turn by a bushfire, you really don't want all your eggs in the one basket, all your shellfish in the one dilly bag, all your seed in the one coolamon or all your wallabies in the one net. People can cope with the losses from a harsh climate because 'they have a broad-based economy, and do not rely on only one or two foods. [This] minimises risks and overcomes shortages of any one type of food much better than can an agricultural community that relies on more restricted food sources. Aboriginal Australia was not vulnerable to famine through the failure of one crop.' But while this is an arguable proposition in relation to the fully fledged agricultural systems—the last 200 years have shown vividly the difficulties of farming in the arid zone—it doesn't help much in relation to the *development* of agriculture.

For one thing it isn't actually true. Presumably one of the reasons agriculture developed in arid areas is precisely because of the uncertainty. One of the principles of agriculture is to smooth over environmental variation in order to keep food supply constant. Hay is cut in summer to feed animals in winter, wheat is harvested and stored, animals are butchered as they mature, and so on. Likewise, arid Australia does appear to use food storage of seeds and dried fruits precisely to tide them over leaner times.

But the proposition that food diversity is a factor is another idea with the germ of a solution, and we will also come back to that.

But the trouble with all of this is that the wrong question is being asked, and it is being asked because two totally separate questions (why didn't Aborigines adopt agriculture from the outside, and why didn't they invent it for themselves?) have been treated as if they were the same question. The result of this is to frame answers about indigenous development along the lines that Aborigines were choosing whether or not to do something, making conscious, economically rationalist analyses of the pros and cons of farming versus hunter-gathering, and deciding that it wasn't for them.

But no people have ever actually *chosen* to become agricul-turalists. 'Okay you guys, listen up, some of us have been having a bit of a chat. We've decided we're sick to death of this unceasing agitation which wears out the animal frame, and is unfriendly to length of days.[5] So we're going to put it to the vote. All those in favour of keeping on hunting? Those in favour of farming? Okay, farming has it. Now, is it to be alpacas or ostriches?' No, agriculture just creeps up on you like old age. One day a young hunter-gatherer, the next an old farmer, but you were never aware of the day you *became* old.

It is a feedback process that builds on itself and locks in people along the way. If you gather in such a way as to maximise yield then you are unconsciously selecting for the factors that improve yield (seed heads that open late, fruit that doesn't readily fall). You can improve yields even more by simple cultivation, along the way perhaps pulling out specimens with smaller fruit or fewer seeds. Once such processes have begun, people become locked into practices that will maintain the crops, because the factors that make them attractive to humans are factors that will make them less naturally viable. The act of maintenance selects for varieties that require even more maintenance. The process with animals is essentially the same—docile animals with big udders and a woolly coat that doesn't shed don't have a big future outside a sheepfold. The two processes converge when the improved animals require improved varieties of plant food.

While these processes are occurring, people who were

originally hunter-gatherers, perhaps paying particular attention to one or two staple plants and animals, are becoming more and more dependent on those plants and animals (which, in turn, are becoming more and more dependent on humans) and less and less able to make use of other species in the normal economy of hunter-gatherers. Increasing density of human population and increasing sedentariness add to the feedback loop. When the cycle is complete, after hundreds, perhaps thousands of years, the people are no longer hunter-gatherers but farmers. Farming is addictive behaviour.

So why not Australia? The embryo of farming was present everywhere, in the millet-harvesting and storing of the inland, the eels of Victoria, the yams of the north, the Solanum of the centre, the fish traps of Queensland, the wallabies of Tasmania. What was it about Australian Aboriginal society that stopped these activities dead in their tracks, and kept them as part of a hunter-gatherer repertoire, before a feedback loop could swing into action with its inevitable conclusion? Well, quite a lot.

First there is kinship, with its obligation to share the proceeds of the hunt. A hunter who is going through a lucky period is giving away some of the spoils as an insurance system, setting up a bank account he can draw on to support his family when he is going through a run of bad luck. The system spreads out of the local groups to surrounding groups through marriage systems, extending the concept to cover whole groups who are going through a bad time. Agriculture would, in theory, make this system redundant, because everyone could be certain of producing food. But it wouldn't be discarded immediately in case of bad times. The result would be that the more successful you were the more you would have to give away, with no prospect of return. Neighbouring groups who were still hunter-gatherers would be constantly skimming off produce from the farmers, but also exerting pressure to maintain the status quo if a group indulged in practices that affected the interests of a neighbouring group. Someone who experimented with the idea of capturing and keeping a mob of wallabies would quickly discover it wasn't a good idea.

Second, there are taboos, another powerful system in hunter-gatherer societies that help to both spread and conserve resources. People by birth have obligations to some species that they may not kill and eat, and conversely rights to use others. People at different stages of life (young, old, female, sick, pregnant) also have rights and restrictions. Restrictions can affect not only the species concerned but also the places in which they can be obtained. No species can be exploited all the year round by the whole community in all parts of its range. Farming, with a whole community exploiting just one or two species, could not function in this social environment. Someone who had the power to maintain a species may have no right to consume it. Many others would be unable to make use of it at certain times, and there would be no alternative available. Farming requires focus, and a taboo system insists on a broad-based economy.

Third, Aboriginal mythology is a view of the universe in which people, animals, plants and landforms are all interconnected, a forerunner of the Gaia idea of the Earth as a living organism. Activities that over-exploit or damage any part of the environment will have disastrous consequences in both mythology and fact. Farming is diametrically opposed to this, requiring concentration on particular species and extensive modification of landforms. Aboriginal mythology would be unable to accommodate the complete harvesting of crops, setting aside large areas for a particular species, moving species across country, clearing land by felling trees, damming watercourses, building permanent fences. Not only couldn't it accommodate a full-blown farming system, but neither could it reconcile the activities required for even quite primitive farming activities.

Fourth, once farming is started, it also needs a system of inheritance, so that gains made in one generation can be built on in succeeding ones; capital in the form of wealth, stock, structures and knowledge needs to be accumulated. Aboriginal society doesn't have an individual inheritance system of that kind, though knowledge can certainly be handed down. It would be possible to develop agriculture in a commune form, but the land ownership system would not lend itself readily even to this.

Fifth, Aboriginal economy could generate surpluses, just like agricultural economies, but the similarity ends there. In agricultural societies, surpluses were used to support people from one harvest to the next. In Aboriginal society, surpluses were used not to provide for small numbers of people for a long time, but for large numbers of people for short times. The surpluses bankrolled the very large gatherings that occurred for ceremonies, gatherings that were of fundamental importance for marriage, kinship obligations, trade, art, music and so on.[6] They were of great importance in maintaining Aboriginal society, and the concept of frittering away a surplus in feeding a family through a winter would have made no sense.

Sixth, while many Australian species could have formed the basis of an indigenous agriculture (as well as the ones mentioned earlier, the concept of kangaroo-farming keeps re-emerging at intervals), there are some things about both the animals themselves and Aboriginal interaction with them that made this unlikely. Early stages of domestication probably involved keeping and raising young animals whose mothers were killed while hunting. This seems not to have happened in Australia. One reason may have been that the need for mobility meant that you didn't want the added burden of carrying or being slowed down by young animals. Another reason may be that it is only worthwhile raising young animals if the effort results in a bigger return when they are mature. This is not intuitively obvious (and may indeed not be the case in Aboriginal circumstances), and in a hunting society you would be better off trying for a bigger animal the next day. More significantly, Aborigines were broad-spectrum hunters, both in terms of species range and age of the prey. It may be that agriculture is more likely to be developed by narrow-spectrum hunters concentrating on only a few species, seeing the potential of young animals, selectively hunting older age groups, and hand-rearing young. A narrow spectrum is something of a luxury permitted by advanced technology. If your technology is such that you cannot guarantee the result of a hunt, then you need to take what you can get, when you can get it. A goanna in the hand is worth a kangaroo in the bush.

Seventh, marsupials had no products that could be obtained from the living animals (though Aborigines made good use of the skins, bones and teeth from dead ones). Three products that were probably important in stimulating domestication elsewhere were milk, horns and wool. Maintaining females (and as a consequence their young) for milking is probably an important factor in domestication. Marsupial teats are hidden in the pouch, and in any case the young remain attached for long periods. Marsupials can only be milked using sophisticated machinery for little return. In the absence of horns it is difficult to evaluate the age and status of individual animals and therefore harder to selectively manage them. The horns themselves may provide an incentive to keep young males. Wool and hair that can be shorn from animals that can then be released to allow the products to grow again is also a stimulus for domestication (including selection for better quality or quantity). The relative lack of predators in Australia means that there was no incentive to try to protect your herds or bring them into camp at night.

It was the combination of these characteristics of the Australian fauna, and, more importantly, the characteristics of Aboriginal society, that prevented the first steps being taken that led to agriculture elsewhere in the world. All aspects of that society are adaptive and strongly interrelated, and none can be readily changed. As a result, the hunter-gatherer configuration remained stable for a very long time here, but this conservatism should be seen as a sign of strength not weakness. In Australia this stability is adaptive, responding to large changes but doing so in a stable way. It was not just that the individual components of the society were adaptive, but that relationships between them were also adaptive, and these relationships provide a self-regulatory mechanism that gives stability. Just as with Tasmania, the absence of agriculture in Aboriginal Australia is not a sign of the intellectual impoverishment of an isolated people but of the primacy of a unique social and cultural system. Neither is it the sign of this people's inferiority or of the inferiority of the animals and plants of the continent.

For a variety of reasons, then, Aborigines didn't farm any of the wildlife of the continent. The concept that you should put a commercial value on animals and farm them in order to conserve them would in any case have been laughable and incomprehensible. A crocodile is part of its ecosystem or it is nothing. A tiger or panda born in a zoo is different only genetically to a sheep born on a farm—you have preserved the DNA (or some fraction of it) but you haven't preserved the behaviour patterns, the interactions with the rest of the species, or with the other animals in the environment. Aborigines knew instinctively that there was no point in 'conserving' individual animals: you had to conserve ecosystems. Farming wildlife was a process that a system established to maintain the Australian environment didn't permit.

In the absence of the development of farming along the lines of the European system in Australia, archaeologists and anthropologists have come up with an alternative—'fire-stick farming'. Was there such a thing, and is it consistent with maintaining the Australian environment?

5

'Opened up a landscape':

Firestick farming and the control burners

What changes have our natives produced in the vegetation of
Australia? It would seem they have produced singularly little
. . . They fitted naturally into the ecology of the land they
inhabited and might have continued to do so indefinitely . . .
Obviously the frequency and extent of bushfires must have
increased greatly since fire-producing man arrived . . . I know
of nothing to suggest that such firing by the Aborigines has
altered the covering of vegetation.

—J. W. Cleland, 1957[1]

Man, setting fire to large areas of his territory . . . probably has
had a significant hand in the moulding of the present
configuration of parts of Australia. Indeed much of the grassland
of Australia could have been brought into being as a result of
his exploitation. Some of the post-climax rainforests may have
been destroyed in favour of invading sclerophyll, as the effects
of his firestick were added to the effects of changing climate in
Early Recent times . . . Perhaps it is correct to assume that
man has had such a profound effect on the distribution of forest
and grassland that true primaeval forest may be far less common
in Australia than is generally realised.

— Norman Tindale, 1959[2]

From my window I see a park-like scene, trees idyllically scattered across a neighbour's paddock. I see a park created by axe and saw, 150 years ago, clearing thousands of trees from an area where Aborigines had lived for 50 000 years. Less than one tree in a hundred, perhaps less than one in a thousand, that was there at the time of white settlement is there now. It is a pattern repeated all over Australia. The hills to the east, mostly uncleared, nevertheless show a pattern of different vegetation types, trees more spaced or closer, undergrowth thicker or thinner. I see this variation as reflecting the bones of the country beneath the trees, the nature of the vegetation depending on such things as slope, and angle to the sun, and parent rock, and soil type and depth, and prevailing wind, and water flow. Just as a gel with electric current passed across it in the laboratory can show the pattern of bands in DNA, so does the vegetation reveal the content of the land beneath its feet. When a fire burns in this country, the way it burns, and the effect it has, depends very much on the vegetation, and indeed on the same factors that influence the vegetation itself.

This is an Australian landscape, and I try to view it with Australian eyes, my body attuned to the rhythms of the country as much as to the movement of blood in the veins. Here is a land where climate and landscape have shaped the vegetation, where climate, landscape and vegetation have created a particular fire regime for this area, and in turn, that fire regime has helped to modify the pattern of vegetation. It is a complex interplay in a land of complex ecology, whose secrets have been hard to unlock. It is a complex interplay whose origins are deep in time, millions of years in time. Astute observers among the first Aborigines 50 000 years ago unlocked the secrets, and learnt to live in this complex land.

Other people have read this landscape quite differently. When they read the reports of the early European explorers, describing a varied landscape and vegetation, they see this variation as being due to the use of fire by Aborigines. Parts where observers used the term 'park-like' to describe the trees are seen as evidence of this, because, they believe, such tree spacing could not have occurred naturally. But fire doesn't create parkland; clearing

creates parkland, and Aborigines weren't in the business of clearing trees.

Why such radically different readings of the same landscape?
There are many reasons, and they are explored at various points
in this book. But I think a primary reason is that those who see
the Australian landscape of 1788 as being man-made are seeing
it through European eyes, just as the first colonists were doing.
In the Northern Hemisphere fires in the forest are rare and are
almost always caused by people. Cool temperate forest tends to
be uniform, pine and birch extending across the landscape,
unchanging forest across an unchanging land. A landscape where
fires burned every summer, regardless of whether humans were
around or not, and where subtle variations in the land created a
patchwork of vegetation, changing frequently in quite small areas,
and whose subtle variations in turn affected fire behaviour, was
as strange to the people of Sydney Cove in 1788 as it apparently
is to many of the people who have written about it. When the
first artists began painting Australian landscapes, they did so with
a European eye, creating scenes that bore little resemblance to
the environment around Sydney, and much to the environment
of Britain and Europe. Artists later learnt to view Australia
through Australian eyes. Scientists and archaeologists must learn
to do the same.

As I write this, Australia, from east Gippsland to Western
Australia, has been burning again. Over the Christmas break I
watched two major bushfires, one to the west of the house, one
to the east. Both were probably started by lightning, as were most
of the dozens of fires around New South Wales. The main
Gippsland fire may have started from an escaped campfire; a few
others were probably deliberately lit.

Australia was burning before the people came, and indeed it
is plausibly suggested that it was the smoke from bushfires that
brought the first Aboriginal people to the northern shores.[3] How
the first arrivals knew Australia was here has always been a bit
of a puzzle, because the water gap is so great that there is no
land on the horizon to guide travellers. It could be that people
came accidentally, blown away in a storm and finishing up on
strange shores, but this seems a flimsy and unlikely way to

colonise a continent. For the colonisation to be purposeful and deliberate and therefore include enough people to ensure success, there needed to be some way to know that Australia was here, waiting for its first people. Bushfires, with columns of smoke extending miles high, would provide that aiming point—the continent was signalling.

But there may be more to it than this. The first people came from a region where fire generally meant people, though it could also mean volcano. It may be that people were aiming at what they thought was a volcanic eruption, but it seems much more likely that they thought they were going to another land that had people. In the wet tropics, the idea that there could be a whole empty continent in which fires burned regularly as part of a natural cycle would have seemed impossible. After the first landing, the first people may well have been perplexed discovering that they had come to an empty land. The fire was on but there was no one home. It was a mistake that would be made again and again by later visitors to these shores, and finally by some archaeologists. The concept that fire means people would have been as firmly ingrained in sailors and colonists from the cold, damp lands of Western Europe as it was in the first settlers from the warm, damp lands of South-east Asia 50 000 years ago.

As I write this, firefighters all over eastern Australia are trying to contain fires started almost entirely by a series of lightning strikes during a period of electrical storms with little accompanying rain, and a very dry late spring. The smoke from the fires is clearly visible from the air, and would be clearly visible from the sea, from the decks of sailing ships full of intrepid explorers, if any were still sailing up the eastern seaboard. When eighteenth- and nineteenth-century explorers and observers write about the smoke from fires, they constantly assumed they were seeing fires lit by humans, and undoubtedly they sometimes were. With the relatively high coastal population, the smoke from campfires, with multiple fires in each camp, would have been numerous and highly visible. But other fires, lit by lightning, would also have often been observed, and not recognised as being of non-human origin. A UFO, cruising along the coast of Australia, say 500 000 years ago, would also have observed many fires. If fire was a sign of intelligent life in a galaxy far away, the aliens would have assumed that Australia was populated, but it wasn't, and they would have found some difficulty in asking the kangaroos how the fires had got started.

In recent years, some archaeologists have seized on pollen records and ethnography and historical records to propose a theory that Aboriginal use of fire greatly altered the Australian environment, a theory further developed and strongly promoted by Tim Flannery. From the moment the Aboriginal people arrived, it is said, fires swept the country with greatly increased frequency, causing massive changes to the vegetation. This remained the case until after 1788, when the removal of many Aborigines from their lands, and the prevention of others from setting fires because of the risk to British property, greatly reduced the frequency of fire. The effect of this, it is said, has been to increase the growth of vegetation and therefore increase the risk of bushfires. It has been a theory seized upon by those who wish to manage the Australian environment by the use of fire in what is called 'control burning'. It is also a theory that suits farming interests wanting to prevent undergrowth and retain grass, and continue the devastating clearance of trees on the grounds that there are 'more trees now than there were 200 years ago'. Partly because it is a simple theory, and appealing, partly because it presents Aboriginal people as 'firestick farmers', and partly because it suits vested interests, it is a theory endlessly presented as fact, not only in the popular media but also in academic literature. People who the public should be able to rely on to cast a critical eye over such theories have simply swallowed and regurgitated it as if it has been engraved in stone and brought down from the mountain. It seems to be the only theory in the social sciences that has been accepted as fact. But it is only a theory, and one with little evidence or substance to support it, particularly when the huge weight of consequences that it carries is taken into account.

The analysis of pollen grains and charcoal fragments has formed a major part of the theory about the effect of Aboriginal use of fire on the Australian environment. The results are presented as if some kind of experiment has been done which proves that the arrival of humans on the continent can be correlated with a change of fire regime. The mistake that has been made is a classic one in science—deciding what the experiment was *after* you have conducted the tests and got the results. The pollen and charcoal graphs are extremely complex and are the result of a complex natural sampling process from the environment to the archaeological site. The graphs produced comprise wriggly lines with many peaks and troughs. There is so much variation in a long pollen sequence that significance could be ascribed to many different spots on it. We also don't know when humans arrived on the continent, nor do we know when they occupied particular parts of the continent. The practice has been to use the pollen record to determine that a particular change at a particular time is the result of human activity, then to use this as proof that humans occupied the land at that particular time. The public are generally unaware that there is no independent test of this. There are no artefacts or other evidence of human arrival in an area when pollen evidence suggests a change has occurred; the theory is simultaneously the proof. The process can lead to absurdities.

When humans did arrive in Australia for the first time as owners and managers of an entire continent, they were given a list of instructions. One of the instructions said, 'keep an eye on the fire, and if it needs maintaining, maintain it'. The ancient Greeks believed that all matter was composed, in different proportions, of four elements: earth, air, fire, and water. They might, had they known of the existence of other continents, have described continents in the same way. Australia, for the last 25 000 years, and at some earlier times too, is composed mostly of earth, fire and air, with precious little water. Other continents generally have more water and consequently less fire and earth.

It is an equation much like the one the firefighters use to assess risk. The fire triangle is fuel, heat, oxygen; the more of any element, the bigger the fire. But the fire needs a spark. In Australia (and elsewhere) there are basically two potential sparks (leaving aside exotic options like volcanoes): humans and lightning. For most of the millions of years before humans arrived, lightning had kept the home fires burning. Did the arrival of man with his firesticks make a difference? A one-word answer is 'no', a two-word answer is 'not really', but there are longer answers still.

Fire is a good servant but a bad master. In my house, in summer, I sniff the air for the faintest smell of smoke as avidly as does any horse or dog or kangaroo. I watch for columns of smoke, visualising again and again how fire could rush up the hill towards us. But if you are philosophical about it, fire is a natural part of the Australian environment and has been for millions of years. Living with the threat of fire in the bush is like living with sharks when diving, or with snakes while walking, or with traffic accidents on a city street. The idea that we should remove every shark from the sea, or every snake from the land, and control-burn to prevent any risk of bushfire is a recipe for even more environmental degradation. As Phil Koperberg, head of the NSW Fire Brigades, said after the Sydney bushfires of 1994, amid calls for massive control burning, 'Do you want to concrete over all the bush?' If you choose to live in the bush, you choose to accept the risk.

It is often claimed that some Australian plants and animals have actually adapted to fire, evidence of an extraordinarily long period (predating human arrival by millions of years) during which fire has been more significant in the Australian environment than it has been on any other continent, but this is probably not strictly true. Many plants have adapted to the environment in ways that also happen to be valuable in times of fire, and animals have adapted to a variety of different habitats, and can therefore survive during different periods of vegetation regrowth after a fire (or after, say, a cyclone, a flood, or just a tree falling in a forest).

A tree that has the ability to regenerate from roots or lower trunk when the upper tree dies as a result of being broken off in a storm, or falls over, rotten to the core, will also be able to respond when the upper part is killed by a fire. Seeds adapted to long hot droughts, and requiring a combination of heat and water for germination, will also find a fire, if followed by rain, a good stimulus for growing new seedlings. There does appear to be evidence that chemicals in smoke can help promote growth in plants, but whether this is a direct adaptation to fire or the incidental effect of some other adaptation is uncertain. But glib statements that Australian plants and animals have adapted to fire should be treated with caution, and indeed it is difficult to see how such an adaptation could evolve.

It is also often claimed that Australian ecosystems have adapted to fire. This is a bit like saying that lawns have adapted to lawn-mowers. All ecosystems, all over the world, have succession. This is what maintains the world environment in a steady state (or did before the irreversible changes, particularly those involving extinction and permanent damage to soils, of this century). The example beloved of ecologists is the tree falling in a forest. When a tree dies and falls, it leaves a space in the canopy through which direct sunlight can come, perhaps for the first time in a century. Additional moisture can also reach the ground without being absorbed by the roots of the tree. Almost immediately, as a result, the thin vegetation of the forest floor will thicken as grasses and herbs and shrubs become vigorous. Tree seedlings, dormant in seeds in the dark conditions, or surviving for a while as stunted seedlings, can also begin to grow. Eventually these will begin to crowd each other out, the floor vegetation will reduce, and finally a single tree, the fastest-growing species, will fill the gap in the canopy again.

More severe disturbance, say from a cyclone, follows the same kind of sequence but on a larger scale. Depending on the climate and whether the disturbance is near the edge or the centre of an ecosystem, and depending on the severity of the cyclone, a series of complete ecosystems, running from grassland through shrub-land, woodland, and back to forest, can be disturbed. If the original ecosystem is a grassland, then the succession may just run through a series of annual grass species, then perennial species and some other non-grass plants being added to finally produce a highly diverse grassland community.

A fire is just another form of disturbance, and its severity will determine where the succession starts from again and how long before it returns to the form it was in before the fire. But this isn't an adaptation to fire as such, nor have Australian ecosystems evolved in such a way as to need fire to keep them in a particular state. Evolution can't work like that.

Here is another triangle. The vegetation of a region depends on soils, climate and topography. The fire regime that can exist in a region depends in turn on the vegetation (though like everything else to do with fire, this is much more complicated than it seems—topography can also influence fire behaviour, and climate can also influence the likelihood of fire by the relationship between how good the growing season is and how severe the drying season is that follows). The actual fire regime in a particular place at a particular time depends on the fire triangle: how hot the weather is, how much fuel there is, how windy it is. Before humans arrived, you could have mapped fire regimes (frequency and strength) across Australia over time. The map would have looked remarkably similar (except that it would have shades of red rather than shades of green) to the vegetation map, which in turn looks remarkably similar to the climatic maps.

Over time, from the ancient Mir satellite, you could have seen the fire regime map change across the Australian landscape, depending on whether the climate was going through a wet or a dry phase. But there would have been little or no change in that fire regime map with the arrival of humans 50 000 years ago. No, no, I won't have that, you say—that doesn't accord with commonsense. Of course there must have been more fire after human arrival. It's just commonsense.

Just as the commonsense view is that Australian plants, animals and ecosystems have adapted to fire, rather like the commonsense view that the sun goes round the earth, there is a body of opinion that believes the Australian environment was created by Aboriginal use of fire over the last 50 000 years. Here is the triumvirate of Norman Tindale, Sylvia Hallam and Rhys Jones, saying superficially different things, but really saying the same thing. The work of these three has in turn been popularised by Tim Flannery.

Rhys Jones said we need to take a hard look at any ecological change in the last 30 000 years and postulate a natural cause only if human causation can be eliminated.[4] Norman Tindale said, 'true primaeval forest may be far less common in Australia than is generally realised'. Sylvia Hallam said, 'these few Aborigines had *opened up a landscape* [my emphasis] in which it was possible for Europeans to move around, to pasture their flocks, to find good soils for agriculture . . . The European communities inherited the possibilities of settlement and land-use from the Aboriginal communities'.[5] She further quotes an 'empire builder' of exactly 150 years ago, J. C. Byrne: 'But as they have been passing from creation they have performed their allotted task; and the fires of the dark child of the forest have cleared the soil, the hills and the valleys of the superabundant scrub and timber that covered the country and presented a bar to its occupation. Now, prepared by the hands of the lowest race in the scale of humanity . . . the soil of these extensive regions is ready to receive the virgin impressions of civilised man.' (It is interesting that, 150 years later, almost identical sentiments are being expressed by those bent on converting native title back to *terra nullius*, and completing the unfinished business of the empire-builders.)

I shall concentrate on Sylvia Hallam here, because the two men get a guernsey elsewhere and her important work has received little attention. The guts of her thesis is this. Aborigines regularly burned the bush. In doing so they kept pushing the ecosystems back to an earlier successional stage, rather in the way that mowing grass maintains it as a lawn. In the case of forests and woodland, such burning had a number of effects. It reduced the number of trees in a given area (presumably Hallam thought this would happen both by killing existing trees and preventing regrowth, though I don't know that this has been spelled out) and stopped a shrub layer developing. As a result there was a greater growth of grass than there would otherwise have been, and the burning maintained this grass in an early successional stage, so it was good for herbivores. The motive for this was that Aborigines would then be more successful in hunting (as a result of the greater number of kangaroos). It is also believed that more useful plants were encouraged to grow.

This idyllic state continued for many thousands of years. When Europeans arrived in 1788 they were stunned to see the 'park-like' Australian environment everywhere, and quickly took advantage of it, with very little effort, to graze sheep and cattle. The irony was that the maintenance of an environment fit for hunter-gatherers had also unknowingly been the maintenance of conditions ripe for invasion by pastoralists.

Now this is a lovely idea, the kind of story that ought to be true. It ought to be true in the same way that the worlds in the movies directed by Stephen Spielberg ought to be true, or those created in books by Terry Pratchett and Douglas Adams. But I have another idea.

When the redcoats landed at Sydney Cove in 1788 they saw what they believed to be wilderness. It was disturbing to live in a wilderness if you had grown up to believe that Nature was meant to be tamed and that the manicured landscapes of England, with long histories of human modification, were the way the environment was meant to be. This unease in the face of an untamed landscape has permeated the Australian psyche ever since. My grandfather, Charles Henry Young, felt it, trying to clear giant trees with axes in the Margaret River area of south-western Western Australia in 1929, as much as any other explorers or farmers before or since. The land needed to be cleared, turned into something like parkland, with grand old spreading trees dotted here and there among lush green grass.

What Tindale, Hallam, Jones and Flannery are saying is that the Australian landscape *was* a managed landscape, created over many thousands of years. But ironically, because both Aboriginal people and their actions had been largely invisible to Europeans, the managed landscape of 1788 has been allowed to turn into wilderness. Paradise has been lost, and the only way to get it back was to do as the Aborigines had done and burn the bush regularly. It is an argument resurrected every time there is a bushfire anywhere. Fire is a part of the fear of the Australian wilderness, and it must be used to control the environment for people, rather in the way people want to remove sharks from the sea.

But there is precious little sign of a lost paradise in the early writings, unless they are read with flame-coloured glasses. Sydney Cove 1788 could and should be the ideal setting for a movie (*Pleistocene Parkland*) directed by Sylvia Hallam.

'*Morning Governor Phillip.*'

'*Morning Captain Tench. I was just thinking what a lovely morning it was to found a new civilisation in the southern seas.*'

'Yes, Governor. And what an extraordinarily good spot. The whole of this world-class harbour already surrounded by parkland.'

'I agree. Look, over there we can put Government House. Just needs an entrance drive for the carriages of guests, and we can have marvellous garden parties on the grass among those widely spaced trees.'

'Yes, and if you look further, there are the Botanic Gardens. Just need a few beds of petunias and they are away with a tourist attraction at little cost.'

'Now we are in the realm of fantasy, Watkin. Who on earth will sail for nine months to see gum trees and koalas? The Japanese? But seriously, we need to feed this motley crew of likely lads and lasses. If you will just arrange for the sheep and cattle to be dropped over there they will find instant pasture and the odd shady tree. Should have lamb chops for that first garden party in no time.'

'Right you are, sir. Isn't it odd that the natives we glimpse occasionally in the distance didn't develop agriculture in an ideal setting like this? Anyone would think they were happy to just prepare the scene for people from a higher civilisation. Maybe in two hundred years' time they will be protesting in that park over there with masses of Aboriginal flags.'

'You are in a droll mood today, Watkin. There will never be another flag raised in Australia but the British flag, I can guarantee that right now. All ashore.'

All fantasy of course (no one would believe that the British flag would really still be flying!). In reality, there was thick forest all around the harbour. The illustrations show this, and the real words of the players (lovingly compiled by John Cobley in his wonderful account of the first year of the colony, a work that should be better known[6]) paint a vivid picture.

7 January 1788 'full of trees which will take some time to clear away.'

January 1788 'it is surprising such large trees should find sufficient nourishment, but the soil between the rocks is good, and the summits of the rocks, as well as the whole country round us, with few exceptions are covered with trees, most of which are so large that the removing them off the ground after they are cut down is the greatest part of the labour . . . there are some parts of this harbour where the trees stand at a considerable distance from each other.'

3 February 1788 'About 4 mile higher than where the ship lay, the country was open and improved the farther we went up and in most places not any underwood—grass very long.'

February 1788 'By the end of the month, the stock which had been landed on the east point of Sydney Cove had eaten all the grass there.'

20 March 1788 'The other escaped through a thick brush which the natives don't like to go into.'

15 April 1788 'we got into an immense wood, the trees of which were very high and large, and a considerable distance apart, with little under or brushwood. The ground was not very good, although it produced a luxuriant coat of a kind of sour grass growing in tufts or bushes, which, at some distance, had the appearance of a meadow land, and might be mistaken for it by superficial examiners.'

17 April 1788 'Mr Ball found all the country he crossed to be a jumble of rocks and thick woods, except one small spot.'

22 April 1788 'we proceeded for a mile or two, through a part well covered with enormous trees, free from underwood. We then reached a thicket of brushwood, which we found so impervious, as to oblige us to return.'

24 April 1788 'We proceeded to trace the river or small arm of the sea. The banks of it were now pleasant, the trees immensely large, and at a considerable distance from each other; and the land around us flat, and rather low, and well covered with the kind of grass just mentioned.'

25 April 1788 'The country around this spot was much clearer of underwood than that which we had passed during the day. The trees around us were immensely large.'

26 April 1788 'well wooded, and covered with long sour grass, growing in tufts.'

April 1788 'The number of sheep which were landed in this country were considerably diminished; they were of necessity placed on ground, and compelled to feed on grass, that had never been exposed to air or sun, and consequently did not agree with them.'

14 May 1788 'The sides of this arm [of the harbour] are formed by gentle slopes, which are green to the water's edge. The trees are small, and grow almost in regular rows, so that, together with the evenness of the land for a considerable extent, it resembles a beautiful park.' (At last!)

28 May 1788 'We then had to ascend a steep rocky hill, thickly covered with brushwood.'

July 1788 'The natural grass was thin, and very few of the sheep brought from the Cape of Good Hope remained.'

23 August 1788 'This, like every other part of the country we have seen, has a very indifferent aspect . . . the coast indeed is very pleasant, and tolerably clear of wood.'

24 August 1788 'The major part of them were sitting in the long grass a little inland.'

28 September 1788 'though the soil is in general a light sandy soil, it is, I believe, as good as what is commonly found near the sea-coast in other parts of the world. The great inconvenience we find is from the rocks and the labour of clearing away the woods which surround us, and which are mostly gum-trees of a very large size . . . It is the rank grass under the trees which has destroyed [the sheep, only one of 70 surviving by now].'

I have quoted from these observations at some length to give an impressionistic view of the environment of Sydney Cove in 1788. It is not the picture of an environment created by firestick farming.[7] The loss of the sheep—not an auspicious start for the world's greatest sheep nation, helplessly watching them die one by one over nine months—must have been heartbreaking. They couldn't survive just on the sour grass—the only ones that did survive were the ones that had supplementary feeding around the tents—so it wasn't a landscape in which good-quality pasture had been created by fire.

Almost everywhere you looked the trees were old growth and there were masses of them, standing shoulder to shoulder like a marine guard. There was thick undergrowth and long, lank grass, not the signs of country recently and systematically burnt. Occasionally they found relatively open areas, along arms of the harbour, or where there were particular soil types, or a particular topography. It may be that some of the small, relatively open areas had been burnt some years previously. Certainly the exploring parties saw fire. They saw trees burning that had clearly been struck by lightning, and they saw trees burning that had been occupied by possums, where Aborigines had lit a fire at the base to smoke them out. They also found abandoned campfires. George Worgan, surgeon on the First Fleet, recorded (on 28 May 1788) a fire burning in 'healthy brushwood'. He couldn't understand why Aborigines would have done this deliberately, if they had, but the 'wind was blowing very fresh today, and perhaps this may favour their designs, if they had any at all, in burning this stuff . . . whenever the wind blows strong, there are a number of these kinds of fires about the country. I have been induced to impute them to accident, from the natives conveying lighted touchwood about the country with them'. He thought 'their being so careful of preserving fire as long as they can seems to imply that the producing of it is a work of great labour to them, for they even carry these lighted sticks in the bottom of their canoes'.

It is surprising, given the firestick farming theory, to find how little emphasis there is on fire around Sydney Cove in the observations of the first few years. Apart from Worgan, there seems to be only Lieutenant Ball, who had remarked (by 29 April 1788) that 'every part of the country, though the most inaccessible and rocky, appeared as if, at certain times of the year, it had been all on fire'. Arthur Phillip confirmed this observation and was later to expand on it when he noted, in February 1790, 'my intentions of turning swine into the woods to breed there have been prevented by the natives so frequently setting fire to the country'. But these are generalisations, and not supported by the observations they stem from. And most significantly, the vegetation that was observed, briefly, in its natural state around Sydney Cove was not vegetation that had been much influenced by fire at all. It had certainly not been burnt in a way required by the firestick farming model.

Untold masses of trees were removed from the colony in the first year. Trees were removed to make living space, but also for use as building materials, roofing materials, fencing, bridges and firewood. Extra carpenters were called for at one point to try to get people out of tents and into buildings, and some massive buildings were constructed. Indeed at one stage, suddenly realising that the clearing was going too far, Phillip declared a reservation: 'The run of water that supplied the settlement was observed to be only a drain from a swamp at the head of it; to protect it, therefore, as much as possible from the sun, an order was given out, forbidding the cutting down of any trees within fifty feet of the run.' It was the first 'National Park' declared in Australia, and symbolically it had to be declared to prevent further environmental degradation as a result of overenthusiastic tree-felling, within two months of the birth of the colony. Every day there seems to have been tree-felling activity, and by the end of the first year, a visitor not realising the extraordinary industry with which the task of clearing had been attacked (even the roots of trees were grubbed out) might well think that the normal environment of this landscape was park-like. It is the first of many examples, both actual and symbolic, where European views of the landscape have masked Aboriginal ones.

But how have these European views come into being?

Once the idea that human use of fire was significant in 'moulding' Australian environments had been suggested, people looked for, and claimed to have found, all kinds of different evidence that it had been the case. Three things were important. First you had to demonstrate that change in vegetation and fire (as measured by charcoal fragments in sediments) had occurred in the distant past, and this meant looking at the pollen and charcoal records found in archaeological and palaeontological excavations. Second, if Aboriginal burning had been an influence, then you would expect that when regular Aboriginal use of fire ended in most areas in the nineteenth century, there should be changes to vegetation and fauna since that time. Finally, to demonstrate the reality of 'firestick farming' it wasn't enough just to say there had been an increasing fire frequency after human arrival on the continent. You had to show that the effects of fire were to the economic advantage of Aborigines, that they were aware of this, and that they acted accordingly. All of these claims have been made; none stand up to closer examination. In addition, although everyone behaves as if the idea of how firestick farming might work is straightforward, in fact two quite different proposals exist. People seem to believe them both quite happily, rather in the spirit of the White Queen of *Through the Looking Glass* believing as many as six impossible things before breakfast.

If the human use of fire was a major factor in changing vegetation patterns in Australia, then we would expect such change to show up in vegetation, beginning almost immediately after human arrival on the continent. Since, as Jo Birdsell has shown, humans could spread all over Australia in a very short time (around 2000 years) after arrival,[8] this change should show up all over the country, matching the first evidence for their presence anywhere (currently believed to be 50 000 years ago). The change would be expected to involve two pieces of evidence as found in archaeological and palaeontological excavations: a difference in the amount of charcoal per year before and after, and a difference in the kind of vegetation in an area before and after. In particular, the vegetation should change from one with fire-susceptible species to one with fire-tolerant species, and the environment should change in the direction of becoming more open and richer in grass. Depending on the region, hypothetical changes might be rainforest to sclerophyll forest, sclerophyll forest to woodland, woodland to grassland. In fact as Tindale suggested, over time, whatever the starting point, vegetation should go through all these phases. Once the change had occurred, it should be irreversible, and stay in place right up to the time Aborigines were moved from an area, or prevented from carrying out their normal burning practices.

There have been many studies that have claimed to demonstrate some aspects of the evidence for the theory. None have demonstrated all elements, and although proponents of firestick farming have generally assumed that you could pick a bit from here and a bit from there in order to construct a complete patchwork quilt, this is not the case. If the theory has validity, then within experimental limitations, all sites should agree.

At the classic pollen site on the Atherton Tableland there was a shift from rainforest to sclerophyll forest at around 38 000BP, right on the then popular date for human arrival in Australia.[9] To the south, though, there was reported to be a change at Lake George at 120 000BP, with the claim that human arrival must have occurred then.[10] Both couldn't be right, of course. Furthermore, rainforest returned at Atherton around 8000BP. At Lake George grass pollen levels are high before 120 000BP, then decrease for a long time until they increase again at 8000BP. Casuarinas, which are assumed to be fire-susceptible (they are not), decrease at 120 000 and eucalypts increase, but Casuarina is again present for 40 000 years from 64 000BP to 22 000BP, right in the middle of the period when human use of fire is supposed to be active. At Atherton, eucalypts and Casuarina both increase at 38 000, but had previously increased at 76 000BP.

This earlier change had occurred without change in charcoal levels, and there are anomalies everywhere with charcoal in the record. At Atherton the amounts of charcoal go up and down in a manner that is hard to equate to either burning by Aborigines or with vegetation patterns. No change in charcoal at 76 000BP, high at 38 000 but low for the very long period 30 000 to 17 000, then high again for 9000 years, low for the last 8000 except for a recent increase. At Lake George, both periods (each 50 000 years long) of abundant Casuarina before 120 000 BP are associated with high charcoal levels, as is the increase in Casuarina between 64 000 and 22 000BP. Another, much shorter sequence in the quite different environment of Kangaroo Island also shows a pattern of vegetation change, charcoal levels and human presence which is difficult to interpret but certainly shows no clear relationship.[11] Whatever the relationship, at best, human presence might be associated with low charcoal level, human absence with high levels. The explanation offered for this is that Aborigines, burning frequently with low-intensity fires, don't generate much charcoal. After they are gone the occasional bushfire generates large amounts. This proposal might be arguable were it not for the fact that high charcoal levels are supposed to be associated with human arrival at Lake George and Atherton.

These sequences are typical of many, and are usually claimed as certain evidence for the effect of Aboriginal use of fire on the Australian environment, and indeed for the early arrival of people on the continent. But if people didn't arrive 120 000 years ago, Lake George certainly doesn't make sense even at this simple level of analysis, and if they did, then the Atherton evidence makes no sense. In fact, it is not a case of one or the other being right, but that neither of them can actually demonstrate human activity because of the anomalies.

The analysis of pollen sequences has been bedevilled by the same philosophical problem as the attitude to fire that has prevailed since 1770. The sequence, as much in the mind of most of the pollen analysts of the 1990s as of the redcoats aboard the *Sirius* in 1788, is 'Aborigines cause fire, fire causes vegetation change'.[12] At least some of the redcoats knew about the effects of lightning. The correct sequence to keep in mind is 'climate change causes vegetation change, vegetation change causes a change in the fire regime, change in fire regime causes vegetation change'. Both humans and lightning are just sources of ignition of fires.

Apart from the serious inconsistencies in the evidence presented for firestick farming, there are two strong pieces of evidence against it.

Australian landscapes form a great bullseye, the colours of the rings running from the red centre through yellows and greens to the blue sea. It is a rainbow of colours, a rainbow signalling alternately the end of rain or the promise of rain. A promise of rain, in Australia, isn't much of a promise. It is the promise of something good coming 'when the boat comes in', or like jam you can have yesterday or tomorrow but never today. The bullseye is formed as a result of rainfall patterns: in dry times the red centre expands, in wet times it shrinks, but the alternating bands stay regularly arranged. Superimposed on this rainfall pattern is a general pattern of temperature variation, one gradient also running from the centre to the sea, but with a second gradient running from north to south. On top of these basic patterns are the local variations caused by differences in topography and soils. Finally, the particular species compositions of the resulting vegetation communities are a reflection of ancient events—movements of continents, movements of species, evolution of new species.

Given all that information, a computer, or a superb botanist, could predict the vegetation at any point on the continent of Australia, or rather could have done before 1788. Conversely the same superb botanist, dismounting from her horse, could have identified her location within a very small radius. (There is a lovely story from geology, where the wonderful William Buckland, an early nineteenth-century geologist, is said to have been lost when travelling from Oxford to London. He dismounted, picked up a handful of soil and said, 'Ah, yes, as I guessed, Ealing'.) You could not do this if the use of fire by Aborigines had had a major or indeed any impact on vegetation patterns. It is also true that each region will have a particular fire regime, depending on the vegetation, with modifications also based on climate and topography, and that a good fire analyst, parachuted blindfold into an area, could very quickly and accurately predict the likely frequency and intensity of fires in that region. That fire regime will occur given a source of ignition, and that source might be either human (deliberate or accidental) or lightning.

But the same proviso, that the expert needs to do the test before 26 January 1788, applies to both party tricks. One of the great mistakes in the firestick farming debate has been to suggest that a giant experiment began in 1788—as if Phillip was carrying a secret instruction about experiments to be carried out on the Australian continent.

'Listen Arthur, Joseph Banks reckons these Aborigines might have had a profound effect on the distribution of forest and grassland.'

'Ah, these romantic scientists, sir, heads in the clouds, no idea of the real world.'

'Yes, yes, Arthur, I know all that, but he keeps putting up this hare-wallaby-brained idea, and he has the ear of the government. So let's put paid to this nonsense once and for all. Just stop those natives setting fire to things—you'll have to anyway or you'll have no corn left—and we'll see whether the vegetation changes or not.'

'But sir, it will change anyway—I've got sheep and cattle and

chickens, I've got dozens of axes and saws to get those lazy convicts moving, I've got a colony to establish and a continent to clear and settle.'

Well, you get the idea. By the time hard hooves have compacted the soil, weeds have sprung up all over, trees have been felled in their thousands, crops have been planted, rabbits are everywhere, as are feral goats, camels, horses and pigs, firebreaks (roads) are cleared, and new sources of ignition are found (fire for burning crops after harvest or burning piles of felled trees, trains, cars, fires of shepherds, fires in chimneys, sparks when electricity wires rub together), we have a whole new ball game in the Australian bush. Much of this change happened well in advance of the settlers as they moved out into the bush, as did the effects of introduced diseases on Aboriginal populations. On the other hand, as fire was increasingly a danger to houses and stock and property and lives, and as firefighting technology improved, fires could be prevented almost entirely in some areas for years. In short, whatever the difference between the fire regime now, almost anywhere in Australia, and that of 210 years ago, it has little if anything to do with the lack of use of fire by Aborigines, and can tell you nothing about the effects of that use of fire.

Science is full of traps for young players. One of the most invidious is the misreading of coincidence for causation. Just because two events or processes occur at the same time doesn't mean that one causes the other. Both may be caused by another (sometimes invisible) process or event, or the two may have nothing to do with each other. The commonly held belief that the extinction of small mammal (and other animal) species is the result of the cessation of Aboriginal burning is one such confusion of coincidence and causation. The arrival of whites on the Australian continent started a series of processes that led to the extinctions: clearing of land, release of feral animals and plants, replacement of native with exotic pastures, changes to water regimes (including dam construction), the presence of hoofed animals, the introduction of disease, hunting with guns.

The effects on the Aboriginal population were equally intense: disease, massacre, removal, land clearance, settlement formation (missions and stations). One of the consequences was that in most areas Aboriginal use of fire ceased, partly because they were forbidden to light fires because of danger to property and stock, partly because in many areas there was no one left still in place on their own land.

But the cessation of fire did not cause extinctions; the two events were just coincident in time. To claim that they did is very dangerous. It could be argued in fact that continuing Aboriginal use of fire on top of all the other pressures on the fauna and flora could have caused even more massive extinctions. My guess is that reducing fire use has in fact protected many of the smaller animals in the few refuges they have left. To start burning extensively again now would be to finish the process begun when the first steel axe sank into the first tree at Sydney Cove in 1788.

The second piece of evidence against the firestick farming theory is Aboriginal diets. This again is apparently opposed to what would seem to be commonsense, but, just as in the case of the sun endlessly cycling around the earth, commonsense can lead you to a conclusion 180 degrees from the right one.

The mythology, among both Aboriginal and non-Aboriginal people, is that Aborigines are big-game hunters. It is a male mythology, as such mythologies usually are, all over the world. The image of the Aboriginal man hunting with his spears and woomera, performing extraordinary feats of strength and accuracy, and returning home to feed the family with a kangaroo slung over his shoulder, is engraved in the psyches of black and white Australians. 'Man the Hunter' is an emotive and evocative phrase, and it is a pity that everywhere in the world, present and past, it was sheer nonsense. Kangaroo and emu hunts were very rarely successful in the past (they tend to be more so these days with guns, hunting dogs, trucks and radios, but there is still great uncertainty). In fact, as we have already seen (see page 54), families were largely fed by women, and largely fed on small items of food. Much of the diet consisted of plant food, fish and shellfish (depending on location), insects, reptiles, birds and small mammals. The mammals eaten in large numbers included rodents, bandicoots, possums and the very small wallabies like rat-kangaroos, potoroos, hare-wallabies and bettongs.

This was the composition of the diet when people first strolled ashore on some northern beach 50 000 years ago, and it remains the diet (supplemented with store foods) even today in some areas. Occasionally, very occasionally, and probably always with great excitement and much male posturing, a larger animal was killed and eaten. But in the remains of campfires all over the country, kangaroo bones are as rare as Australian Winter Olympic medals. It is one of the reasons why the idea of megafaunal extinctions being caused by overkill would have been cause for laughter around any one of those campfires, any time in the past 50 000 years.

So what has this got to do with fire, other than as a place to roast goannas? Well, just this. It is the kangaroos and large wallabies that do well out of fire. After the fire has passed and the rains have come (if the rains have come), up springs green grass, and the big macropods come in to eat it. But fire is rarely good news for the small animals, who need, as a major component of their well-being, the shelter that only thick shrubs, big tussocks, old logs and leaf litter can provide. If you were to regularly burn to the extent suggested by the firestick farmer theorists and the control-burning zealots, the small species in the ecosystem would rapidly become extinct.

Firestick farming devotees believe that the motive for the use of fire was to improve the conditions for the big macropods (though they also believe, rather in the spirit of believing two contradictory things before breakfast, that fire was bad news for the even bigger macropods and other megafauna). It isn't hard to imagine the reaction around the campfire when this plan was announced.

'Listen, me and the boys have been talking, and we reckon what we should do is a lot more burning.'

'Why is that, dear?'

'Well, you know how you and the kids like your kangaroo steak every night? Well, me and the boys reckon this is the way to get it. What's more, we wouldn't have to move, just light a fire, sit here, wait for the grass to grow, and in come those big old man kangaroos.'

'And, er, what are we going to eat while we wait for the grass to grow?'

'Um . . .'

'And just remind me, will you—the last kangaroo you and the boys got was what, three months ago?'

'Well, only twelve weeks, actually.'

'And since then we've been eating all these small animals that live in this old-growth woodland, right? Those same animals who will be driven away by fire and take ten years to come back, assuming that our neighbours haven't had the same hare-brained scheme and burnt their woodland and there are some animals to come back.'

It doesn't make sense, either as a deliberate strategy or as an accidental outcome. You wouldn't set out to damage your main food source by setting fire to the supermarket in order to roast a crocodile steak. If it had been done accidentally, the result would have been the extinction of masses of small species, on a scale even greater than that seen in the last 200 years. The argument by people who haven't fully understood the firestick farming theory is that the small species have adapted to later successional stages, and that fire is an advantage because it removes all the accumulated debris and starts the cycle all over again, so that small species can increase their abundance as conditions favourable to them develop. You can only put this forward as a proposition if you are looking at a single area in isolation.

In recent times Tim Flannery has also had a go at the question of 'motivation', having accepted all of the other propositions put forward by Tindale, Jones and Hallam. Flannery's idea of the reason for 'firestick farming' seems to boil down to this. What if humans arrived on the continent and wiped out the megafauna instantly, and the effect of this was to change the habitat because the megafauna were no longer there to eat the grass? This would be a problem because the vegetation would become thicker and we know that this wasn't good for the Aboriginal economy, and we also know it didn't happen because everyone knows the vegetation was open and park-like. So Aborigines were forced to use fire, to keep regularly burning, in order to compensate for the loss of the megafauna. What was going on was that they had to keep burning and burning, but really they were just running very hard to stay in the same spot. When they stopped the vegetation grew back again, so we have to take up the torch that was dropped, and assume the black man's burden of regularly burning the bush because we don't have Diprotodons any more.

As Father Brown might have said, 'No point in searching for a motive, my dear Flambeau, if no crime has been committed'. There is no 'crime' here. Flannery presented no new evidence for the reality of firestick farming, so trying to work out why something might have happened that didn't happen is futile. But leaving this aside for the moment, the motive being suggested here makes as little sense as the motive others have proposed.

Flannery is forced to adopt Paul Martin's 'blitzkrieg' extinction model for the megafauna. There are two reasons for this. If the megafauna didn't become extinct immediately humans arrived, then there was some very long period when human use of fire modified the environment before this event. And you would have to suggest that when the extinctions did occur, whatever the use of fire before, it had to be different after. Unfortunately, there is in any case no evidence for Aborigines causing megafaunal extinctions, let alone causing them 50 000 years ago, and I shall examine the causes and timing of extinctions in more detail in Chapter 6.

But why should the loss of megafauna, whenever it occurred, affect the vegetation anyway? Even if the extinctions were instantaneous and synchronous, the effect of the loss of large grazing animals would be short-term and seasonal at most, and the slack, if it existed, would be taken up by increasing populations of the smaller macropods that survived, and even by groups such as insects. While the idea of blaming fire on extinctions instead of extinctions on fire is ingenious, it doesn't solve anything, and runs into the brick wall of time of extinctions if nothing else. Whatever your proposal about past events or future practice, if you're relying on firestick farming for support you're in trouble.

The firestick farming idea has two different, and contradictory, models. One seems to be this. A band of Aborigines observe that part of their country hasn't had a fire through it for some years.

They decide to burn it. In doing so they are careful not to let their whole country go up in smoke, and as a result parts of their country remain unburnt. The part that is burnt is cleared of leaf litter, fallen branches, logs, shrubs and long grass. But after some interval grass has grown and kangaroos have come in to feed and can be hunted. A year or two later shrubs and useful plants are starting to regenerate, and some other species of animals come in who have adapted to this successional stage. A few more years and other species come in, and so on until the particular piece of country has returned to its original state. Then it is burnt again. The periods involved have not been specified in detail, but it is obvious that even old-growth grassland would take many years to return to its former condition, while some forests can take more than a century.

So what next? Well, while the grass is growing in the first patch the band might set fire to another area, then later on another, and so on. That is, throughout their territory they would have a kind of patchwork quilt, each with a different crop, of vegetation in every different stage of regrowth. Each patch can be reseeded and restocked at different times by the appropriate blend of seeds and animals from other patches at different regeneration stages.

This idyllic picture, like the patchwork of different greens and browns we see in British tourism posters, is very appealing. It would represent environmental engineering and management on an astonishing scale. But the doubts about it begin to surface as soon as you start to explore the implications.

How big is a band's territory? Obviously it varied enormously in different parts of Australia. Rhys Jones, coiner of the term 'firestick farming', thought an appropriate figure might be 30 km² in well-populated areas, so we will stick with this as an order of magnitude to guide our mind's eye. This is 900 square kilometres. If we say that full regeneration takes a hundred years, then each patch that is burnt can only be 3 km². If full regeneration time is less, then the area can be slightly bigger. How on earth could you control burning to this extent? Today, with the full armoury of modern firefighting techniques and equipment, and in country where trees have been thinned, and grass stripped bare by stock, even minor fires can run for kilometres. What we are talking about here is burning mature habitats, with all the combustible material that goes with them.

And what of lighting strikes causing fire before you are ready to re-burn? Or a fire from a neighbouring band getting away? The areas are not fenced off with fireproof fencing. Any accidental burning on this model is going to extensively cut down the proportion of habitat in the later stages, and therefore reduce the chances of having reservoirs of species to restock. And what on earth is it for? Why, if the aim is to increase grasses and habitat for kangaroos, would you set up a system that gave you only 1 or 2 per cent (depending on regeneration time) of your country in this apparently desirable form? And how would you keep the system going, with the necessary precision, over extended periods through shifting band composition and territory in a society without written records?

What is more, although such a system is implied in much of the writing about firestick farming, and is probably the popular view of what it is all about, it isn't actually what is intended. It is a model (if it actually worked!) for environmental stasis. The aim is to have every part of the band territory cycling through the same stages, and whenever you looked at it (like a steady-state universe) it would look on average the same. This isn't a model that creates woodlands out of forests, grasslands out of woodlands. This isn't in fact a model that creates a park-like landscape for the hands of industry to unload their sheep down the gangplank of the ship into. So let us look at firestick farming model number two.

In this model the aim is to 'open up the country' by the use of fire in a way that anticipated the use of axes 50 000 years before 1788. Because the very early successional stages after fire, it is said, are the most productive ones for the Aboriginal economy, what we need to do is keep regularly burning, on the same principle as regular lawn-mowing. This does two things, though one is never spelt out. First, it removes litter, burns large tussocks and old grass, and removes shrubs, so that we have clear ground in which grass can grow afresh. Second, it prevents the regeneration of trees from seedlings, so that as trees die (or, presumably, are sometimes killed by fire) they are not replaced.

In this way forests are thinned and woodlands are opened. But the mechanism for the loss of tree numbers is not spelt out and is contradicted elsewhere. The most that has been said seems to be that one kind of tree may replace another kind, and there is not even any good evidence for this process resulting from fire, let alone Aboriginal use of fire.

But in any case, you can't change vegetation patterns by burning at long intervals. You need to do it regularly at short intervals, and you need to do it over your whole territory. One of the popular myths about firestick farming confirms this view. The propaganda suggests that by regularly burning, preventing the build-up of litter and vegetation and thick trees, you prevent wildfires. Each time you burn can be a 'cool' burn, very gentle and controllable, and extremely environmentally friendly. So you need to keep all your country at this early successional stage, and so do all your neighbours.

Rhys Jones 'confirms' this model with some figures. At one time he suggested that a 'significant fraction of the continent would have been burnt once a year' and as a minimum 'few regions would have escaped fire for more than a decade or two'. Later he came up with an estimate of '5000 separate bush fires being lit within each band territory of 900 square kilometres each year'. That is, about five fires per year per square kilometre. Some people don't mow as frequently as that.

If this is the model, the extinctions of small animals would have been massive. If country never reaches a successional stage of more than a year or so, and if everyone is doing this, there would be no reservoirs for plant and animal species to recolonise, and no opportunity for them to do so. Erosion would be severe as the surface of the ground would be constantly cleared, and the dust would have been blowing. There would probably have been major changes in vegetation patterns across Australia, with the continent having a distribution of fauna and flora directly related to the distribution of humans.

This is the kind of model that foresters tend to have in mind, and it is the kind of approach called for after every Sydney bushfire. It has been the European approach to the Australian landscape since the first anchor buried itself in Sydney Harbour, but it is not the approach that had been used for the 50 000 years before the rattle of that chain was heard. The landscape didn't look like this when Arthur Phillip surveyed his new domain (and Domain), and the sediments tell us that it didn't look like this in the past.

Many years ago I warned that firestick farming was in danger of becoming a self-fulfilling prophecy, and my prophecy has proved true, as Flannery, among others, has shown. Biological and archaeological data are not being treated as a test of what is only a hypothesis, but rather the data is being treated with the assumption that firestick farming is a reality. Interpretations made on this basis are then, in a process of circular argument, seen as providing further evidence for the hypothesis. Rhys Jones gave this process his blessing when he suggested that we should take a hard look at any ecological change in the last 50 000 years and postulate a natural process only when interference by man can be discounted—a case of guilty until proven innocent. On the contrary, we should take a hard look at any ecological change in the last 50 000 years and postulate interference by man only when natural processes can be discounted. If every change in vegetation seen in the pollen record, and every patch of open vegetation, and every extinction of an animal species, is simply treated as further evidence of firestick farming, there is no chance of establishing the truth. Instead, changes should be analysed carefully to see if a cause can be established.

Theories that humans caused the extinction of the megafauna are in large part a reaction against the idea of hunter-gatherers being in harmony with Nature. The firestick farming hypothesis is a result of the same cause; as Rhys Jones said, both are 'striking examples of the power of hunting and gathering man to alter his environment'. The general approach seems to be that people of ill will believe hunter-gatherers are second-class citizens in comparison to farmers, and people of goodwill should therefore see them as being more like farmers.

But the difference between farmers and hunter-gatherers has to do with process, not ranking, for they have fundamentally different ways of interacting with the environment, and there is no exchange mechanism for conversion between the two currencies. The difference is between 'interferers' and 'observers'. Farmers interfere with climate (through irrigation), soils and succession in order to successfully introduce foreign plants and animals into an area. The systems established are unstable and have to be maintained by constant interference. Hunter-gatherers are observers, relying on detailed knowledge of climate and native plants and animals in order to extract energy from a system. The system remains stable because the observations are designed to detect surplus, and there is no interference in the process that generates the surplus.

It has long been obvious that hunter-gatherers have a detailed knowledge of the behaviour and ecology of individual species of plants and animals. It is less obvious that they also need to have a detailed knowledge of community ecology. The characteristics, both seasonal and longer term, of different habitats need to be known in order to harvest a range of resources efficiently. In Australia, a fiery continent since the beginning of time, the knowledge needed to include an understanding of the long-term effects of fire. Come back one year after a fire and there will be kangaroos and many plants with fruit; two years after a fire wattles are starting to produce seed and native rodents will be abundant; after five to ten years some small wallabies are available and wattles are in full production; and after twenty years small wallabies, possums and bandicoots are all abundant.

At this climax stage, fire (or flood or violent storm) may start the whole process again. The fire may be started by lightning, or an escaped camp or hunting fire, or may have occasionally been set deliberately to clear up bush if accidental fires hadn't happened first. A casual attitude to fire was possible because the regeneration sequence was known. But a deliberate attempt to interfere with the system would have been disastrous. Frequent burning would have prevented the development not only of the plants and animals of later successional changes themselves, but of their role in soil decomposition systems and nutrient status such as to seriously affect even the early stages. Either you farm or you don't. You can't half farm, and firestick farming would be half farming.

There is, however, a natural fire regime in Australia that Aborigines were not only aware of but made use of, their use of fire providing an alternative source of ignition to lightning strikes. Fire has for millions of years been as much a part of the Australian scene as drought and flood, and the interacting cycles of these forces with vegetation took little notice of the arrival of humans 50 000 years ago. It would be bitterly ironic if a mistaken belief that Aborigines used fire extensively and damagingly caused us to do the same, but even more damagingly. With the Australian population much greater and the trees much fewer than they were 200 years ago, we have little room to move.

Yesterday I walked down to the stream below my house. I carried a mattock to remove some thistles, British invaders in bright colours who have followed the watercourse and established colonies of prickly individuals. I battled away for some hours, feeling despondent because there were so many of them, and every one I knocked down had many offspring. Coming back up the hill, though, refreshed my spirits. To the south are bare paddocks of introduced grasses, eaten down to near bare ground in a dry year, but I walked up the hill through climax grassland. My hill has not been burnt for many years, nor, thanks to Vicki McLean, our predecessor on the property, has it been overstocked—the two effects are much the same. It is a complex of huge old Poa and other tussocks, interspersed with other native grasses, and a range of lichens, herbs and small shrubs. It is in fact old-growth grassland.

Just as we need to protect old-growth forest as a unique and irreplaceable highly diverse climax vegetation form in some areas, so too we need old-growth grasslands and woodlands and deserts and coastal communities. We won't keep them with a regular program of 'control burning'.

Wilderness isn't 'jungle', it is landscape that has reached its potential without human interference. If every part of Australia is subject to human modification, we are headed for environmental disaster. There is so much now that has been modified that if the view that the Australian environment was created by Aborigines prevails, then the Trojan Diprotodon will be used to get rid of all wilderness.

The colonists, in putting forward a view that Australia was a *terra nullius*, thought they were ensuring that all of this land was up for grabs by people of British descent protected by the military. The Mabo decision in the High Court caused outrage among their descendants, and the Ten Point Plan was designed to ensure that the error of not getting rid of all vestiges of Aboriginal ownership is reversed once and for all. To deny the existence of several hundred thousand people in 1788 was bad enough. To repeat the denial in 1998 was obscene.

The debate about wilderness has taken the same form (and has been played out among much the same protagonists). The original colonists saw a whole continent of wilderness, there to be subdued, tamed, domesticated for human purposes into a managed landscape, designed as a kind of factory for short-term profit-making.

The Australian environment has also been a place for prophet-making, with environmentalists managing briefly to have some areas declared off limits to human interference. Now, with the New Right component of the environmentalists holding as much sway as their counterparts have had in economics, the process that began with the first axe hitting the first tree in Sydney Cove in 1788 will be completed.

So what about the Diprotodon and other megafauna? If you can't blame fire for causing extinctions, and you can't blame the extinctions for causing fire, can you blame people for causing the extinctions anyway, by some other mechanism?

6

'The extinction of such pachyderms':

The great megafauna debate

It seems to me that the conditions of life can have very little changed [in western New South Wales], as the same shells live still in similar waterholes. The want of food can scarcely be the cause of their [Diprotodon] disappearing; as flocks of sheep and cattle depasture over their fossil remains. But as such a herbivore must have required a large body of water for his sustenance, the drainage of these plains or the failing of these springs . . . has been, probably, the cause of their retiring to more favourable localities.

—LUDWIG LEICHHARDT, 1844[1]

This is known as the Pleistocene overkill. Sentimentalists among us still try to insist that it was a changing climate, not mankind, that did the damage, or that we only delivered the *coup de grâce* to species that were already in decline. It is remarkable how strong remains the wishful thinking for finding an excuse to believe in climatic change.

—MATT RIDLEY, 1997[2]

It's an old debate, and like many old debates, a good one. What caused the extinction of the Australian megafauna in the late

Pleistocene—human action or climatic change? It is reminiscent of the old debate in physics about whether light was a wave or a particle, and I remember the imaginary dialogue that answered this. 'Is light a wave?' 'Yes.' 'Is light a particle?' 'Yes.' Some people have tried to create the same dialogue for megafauna extinction. 'Did human action cause the extinctions?' 'Yes.' 'Did climatic change cause the extinctions?' 'Yes.'

But this is a cop-out, and no answer at all. The real way to frame the question, as is the case with so many other questions about Aboriginal impact on the Australian environment, was spelled out brilliantly by Rhys Jones. 'If man had not managed to cross the last water channel of Wallacea those distant tens of millennia ago,' asked Rhys, 'would our knowledge of the late Pleistocene "giant marsupial" fauna only have come to us from the bone breccia of a Wellington Cave—or would at least some large beasts, lumbering down to the water's edge, have graced the sketch books of a Joseph Banks or a Charles Lesueur?'[3] In short, would the megafauna have become extinct had humans not been on the continent? People who have addressed this question have probably been roughly equally divided in their answer, though fashion swings the proportions one way or another over time.

In the early days of Australia the landscape must have seemed as alien to the British and European colonists as the landscapes of Mars and Venus do to us today. Such harshness of view, such strange and rough plants and animals, such monstrous swings in climate. This view of Australia coloured the Australian psyche generally, but had, in the context that concerns us here, a particular effect on the scientists who were investigating the past. Not only did this environment make archaeology a very difficult endeavour, but it also left little doubt in the scientific mind that environmental change was the most logical explanation for the extinction of the giant marsupials.

Something else that set the tone for theories was the extinction of the giant mammals of Europe. There was a pretty clear picture of a quite different world in Europe, where glaciers covered the continent, people lived in caves for warmth, and woolly mammoth and rhinoceroses roamed a snow-covered landscape. That environment had clearly gone, and with it the animals, and that was a fairly clear-cut link. In 1890 William Anderson made the link between mechanisms of extinction in Europe and in Australia explicit: 'there can be little doubt that [glacial conditions] were among the most potent causes of the extinction of these large Pleistocene vertebrates, just as has been proved to have been the case in other parts of the world where the large Pleistocene forms of life died off on the advent of glacial conditions of climate.'[4] In Australia, although there was some evidence for glacial action, it soon became clear that drought and aridity were considerably more prevalent than snow and ice.

From Ludwig Leichhardt onwards, himself soon facing his own extinction (either because of climate or by human action), observers lined up to link climatic change to various mechanisms of extinction. One author, for example, in a grand sweep in 1879, spoke of an early wet climatic period when

> large mammals roved over the land . . . Conditions were favourable to the growth of succulent herbage capable of sustaining a large and varied mammalian fauna . . . This period was brought to a close by the lowering of the land . . . The fertile tracts of the lowlands were submerged, and the productive powers of other areas would be diminished by the gradual desiccation going on . . . a struggle for existence would ensue, in which the less adapted and less easily modifiable would succumb.[5]

Not only the past but the present gave evidence for those who would see. Another in 1885, observing one of the many severe droughts of the nineteenth and other centuries, noted:

> striking evidence has been afforded by the dryness of the last few seasons, of how quickly, through want of rain, and overstocking, savannahs of waving grass may be converted into desert-like plains; and the immediate influence of these climatal changes on the fauna is fully attested by the numbers of kangaroos and emus, which died last year through want of sustenance on the western plains.[6]

Having seen what drought could do in the 1880s, he applied the lessons to the marvellous Cuddie Springs site, which would continue to be investigated for over a century. There had been water, as the remains of crocodiles showed, 'in abundance' and then it had gone. Bones of megafauna were there in abundance too, and

> nothing but want of water could have brought together such a heterogeneous assemblage of animals to the same drinking place; and what must have been their last terrible struggle for existence, as the supply of water failed, must be beyond description. This one instance may be taken as typical of the general cause of the disappearance of these animals since Pleistocene times . . . Diminished rainfall . . . probably led to the gradual dying out of the once rich Pleistocene fauna.[7]

Modern observation and ancient fossil also combined to good effect at another classic site, Lake Calabonna in South Australia. Thousands of giant marsupials and birds were there, bones not jumbled up together as at Cuddie Springs, but in the form of complete whole skeletons. They had therefore died where they were found, and

> met their death by being entombed in the effort to reach food or water, just as even now happens in dry seasons, to hundreds of cattle which, exhausted by thirst and starvation, are unable to extricate themselves from the boggy places that they have entered in pursuit either of water or of the little green herbage due to its presence. The accumulation of so many bodies in one locality points to the fact of their assemblage around one of the last remaining oases in the region of desiccation which succeeded an antecedent condition of plenteous rains and abundant waters.[8]

There was no evidence of human presence at either Lake Calabonna or Cuddie Springs (though evidence would be found over a century later in a different part of the Cuddie Springs site, using the more advanced archaeological techniques available), and if you were a gambler you'd have had your money very firmly on climate change as the cause of megafaunal extinction. But if it was as simple as that, the debate wouldn't have raged for 150 years. Running alongside these environmental explanations was a very strong true believer in human causation. And a very powerful person he was too: Richard Owen, curator at the British Museum, and a towering figure in the scientific establishment of Victorian England. What was his problem with theories of climatic change? Well, it lay in the potent phrase 'struggle for existence'. Where did a struggle for existence lead you from 1859 onwards? It led you to the theory of evolution, and Owen wasn't having a bar of it. As in the late twentieth century, megafaunal extinctions were a powerful piece of evidence in relation to religious or quasi-religious beliefs. Why was Owen so opposed to the theory of evolution? It may have been religious beliefs, but it was more likely to be jealousy—Darwin, someone who Owen probably saw as extremely junior, had created this work that had shaken the world, and it had to be stopped by the most senior biologist. It is not an uncommon reaction of eminent thinkers to great works someone else has created.

But the religious climate was also important, of course, and the mid-nineteenth century was a great battleground between science and religion, in Australia as elsewhere.[9] Darwin's theories generally received a hostile reaction in conservative Australia; there was a massive assault on Darwinism from pulpit and lectern, with people flocking forward to be counted on the side of the angels. They included people who Darwin might have confidently expected to be on his side, but many intellectuals and scientists were shocked by the implications of *The Origin of Species*.

Owen had toyed with evolutionary ideas before 1859, but as the wave broke and he found himself left behind by the new biology, he became Darwin's most implacable opponent. Owen's attitude to evolution coloured his attitude to the question of the extinction of the Australian megafauna. In 1843, just four years after the Australian megafauna had first been discovered when Thomas Mitchell's party descended into the Wellington Caves, Owen's fertile and intelligent brain was at work on the implications. And what did he come up with? Why, climatic change of course:

time was when Australia's arid plains were trodden by the hoofs
of [Diprotodon]; but could the land then have been, as now,
parched by long continued droughts, with dry river courses
containing here and there a pond of water? . . . May not a
change from a more humid climate to the present peculiarly dry
one have been the cause, or chief cause, of the extinction?[10]

But then came 1859, and Darwin observed that Owen's
review of *The Origin of Species* was 'extremely malignant' and 'full
of spite'. Owen in fact was trying to have it both ways, attacking
Darwin's theory while simultaneously claiming to have thought
of it first. Certainly he was not going to advance the cause of
natural selection, and continuing to present megafaunal extinctions as being caused by climatic change would have done just
that (remember the 'struggle for existence' at drying waterholes).

In a paper published in 1870, Owen had two bob each way,
speaking of the extinctions as 'exemplification of the fruitful and
instructive principle which under the phrases "contest for existence" and "battle of life", embodies the several circumstances,
such as seasonal extremes, generative power, introduction of
enemies etc, under the influence of which a large and conspicuous
quadruped is starved out, or falls a prey, while the smaller ones
migrate, multiply, conceal themselves and escape'. But he went
on to speculate about the 'introduction of the Human kind' and
'the final extinction'. In 1877 he was using ethnography: 'as the
elephant succumbs to the spears and pitfalls of the negro hunters,
the minor bulk of the Diprotodon is not likely to have availed
it against the combined assaults of the tribe of Australoid wielders
of club and throwing sticks.' By 1879 he had a mechanism for
the selective extinction of the large animals: 'to a race of man
depending like the blackfellows for subsistence on the chase, the
largest and most conspicuous kinds of wild beasts first fall prey.'[11]

So for Owen there were no evolutionary implications in the
extinction of the megafauna. But he would have been whistling
in the wind about this except for one thing. The proponents of
a climatic mechanism had failed to come up with a mechanism
by which climatic change would cause the extinction of *some*
species and not others. In fact, the examples they used at the

time, of cattle and kangaroos dying in drought, referred to animals smaller than the megafauna, and in the case of kangaroos, to species that had survived the extinction period quite happily and continued to do so. Owen, on the other hand, did have a mechanism—hunters preferred bigger animals and had hunted them to extinction first.

There was another reason for the continuation of the debate, and it too had psychological overtones. Archaeological investigation into Australia's past, and the attempt to find out whether humans had been here a long time, eventually came to rest, in the absence of radiocarbon dating, on finding evidence of association between humans and megafauna. The only really convincing association, in a land of shifting sand dunes and deep cracks in the earth, was going to be evidence of human modification of megafaunal bone. It was the only proof about which there could be no argument of association. A scientific community searching for cut marks on bones did not want to be told that climatic change had caused extinctions. If that was the case, then there may be no 'kill sites' with Diprotodon bones chopped with stone axes, which would have proved ancient human occupation.

In spite of Richard Owen, the scientific community in Britain fairly quickly came to accept the obvious (once pointed out!) truth of evolutionary theory. Australia was a much harder nut to crack, where some scientists (in an eerie parallel to the views of conservatives today about the world as a whole) had 'never cause to entertain any doubt, that we are surrounded by species, clearly defined in nature, all perfect in their organisation, all destined to fill by unalterable laws those designs for which the power of our creating God called them into existence'.[12] Even as late as 1876, William Macleay, then president of the Linnean Society of New South Wales and therefore a leading member of the scientific establishment, had to 'admit' that the 'testimony of the rocks, so far from giving ground for a theory of continuous modification of form, seems rather to afford proof that there may have been many successions of distinct creations at long intervening periods'.[13] It was really not until after 1900 that opinion turned in favour of Darwin, by then safely dead.

So the debate ground to a halt. You could document climatic change all you liked, but without a mechanism that selectively removed large species you were wasting your time. You could postulate club-wielding 'Australoids' sweeping across Australia like a horde of vandals, but without the kill sites the evidence was only circumstantial. It was, in fact, very difficult to put people at the scene of the crime. Nobody had much doubt that human occupation was very old, and that humans had seen megafauna alive and well, but demonstrating this proved devilish hard. Strangely, radiocarbon dating, shining a light into every other nook and cranny of Australian prehistory, failed to shed much light on this question.

In the nineteenth century, belief in the causal agent in the extinctions depended on where you stood in relation to religious beliefs, the evolution debate, and your vested interest in finding an old archaeological site. The absence of a mechanism on one side, or a site full of butchered Diprotodons on the other, meant that faith and belief were enough.

In a sense, not much has changed as we wearily reach the end of another century, but the current scorecard would probably see something of a majority in favour of human causation. Why is this so? Well, there are several vested interests with bets on the outcome of this debate. Some argue for the proposition that Aboriginal people do have a right to the country because they did modify the environment massively, others that if Aborigines modified the environment massively so could, or should, we. Others want to argue against any notion of Aborigines as conservationists, acting in the interest of society, because they want to promote the idea that greed is good, and that the natural state of humanity is all for one and all for one—good government is no government, and we should only conserve things, by private ownership or commercial development, that are of value to humans, who were after all put on earth to hold dominion over the birds of the air and the beasts of the field. Furthermore, if climatic change caused such massive extinctions, then we really should get our act together on greenhouse gases pretty quickly.

As evidenced later, that is not a message that big business and its allies want promoted.

How has this conglomeration of interests managed to tip the weight of the debate to their side? Has new evidence been found? Not really, but it hasn't seemed to matter. In 1967 Paul Martin, from Arizona, argued in a very influential book the case for human causation.[14] 'Pleistocene Overkill' was the evocative term. It was little different to anything that had come before except for one absolutely crucial and magnificently simple proposition. The lack of kill sites had always been a problem. No problem, said Martin. The essence of overkill is that it happens very fast, so fast in fact that there are very few kill sites and your chances of finding them are correspondingly low. In fact the faster the extinction the fewer the sites, so it could be argued, and was, that the more you looked and failed to find kill sites the better the evidence for overkill. Finding sites with evidence would have been contrary to the interests of the model. It was marvellously religious in its overtones, requiring great faith, and the greater the faith the better you were. Or, as Joseph Heller suggested in a slightly different context—Catch-22.

In America the game was easy. Humans hadn't been there very long, only around 12 000 years, and the American megafauna had become extinct some time around then. Not much time to play with anyway, and there was no rule about how fast 'fast' had to be when it came to overkill. In Australia it wasn't so easy. Radiocarbon in the hands of people enthusiastically looking for older and older sites had pushed Australia's occupation at least back to 40 000 years (subsequent work with newer techniques would hit what is probably about 50 000, the figure I am using). Conversely, sites with megafauna that looked to be a lot younger than this kept popping up everywhere. The problem was that there were an awful lot of years to play with, and even if you only had megafauna surviving to, say, 26 000 years ago (as at Lancefield), it meant humans had coexisted happily with megafauna for at least 14 000 years, and it would be a bit odd if they had suddenly turned around and started massively butchering them after that length of time. Similarly, it would be odd if the megafauna had suddenly decided to make themselves available to be butchered after 14 000 years of

watching humans in action. Paul Martin's big idea needed naive
animals and inexperienced people in a short, sharp, explosive
mixture, like a shoot-out in a saloon where bullets fly for a few
minutes and then the smoke clears to reveal bodies all over the
floor and the lone sheriff still standing. It had to happen fast or it
wouldn't happen.

What to do to save the theory? Well, Rhys Jones rode in
with the solution in the manner of a circuit judge declaring all
the prisoners innocent. None of the dates for megafauna younger
than the time of human arrival were valid, he said, therefore
extinction had happened quickly. Hang on a minute, your worship, you feel like saying, aren't you assuming what you are meant
to prove? No, says Jones, like the Queen of Hearts in *Alice in
Wonderland*, sentence first, trial later. It is a technique that continues to be used successfully by Tim Flannery. But what if there
were a plausible mechanism for climatic change acting to select
only the megafauna? Wouldn't that tip the balance back again?

The Pleistocene is a classic example of the kind of 'unknown
regions' that 'should be preserved as hunting-grounds for the
poetic imagination'. So let us see where a poetic imagination
might take us in a voyage back to an unimaginably distant time.
What do we see as we step out of Dr Who's Tardis? Well, if we
have stopped anywhere up to 10 000 or 12 000 years ago, we
would probably think, like the time traveller in H. G. Wells's
book, that we hadn't moved at all. There would be the same
trees, the same kangaroos, the same birds, the same ground, the
same feeling that here is a continent where there is just enough
water to survive. Australia is a continent on survival rations.

Go back just a bit further. If you thought Australia was a dry
country in the late twentieth century, take a look at it now. This
is the worst nightmare of Greenhouse theorists or El Niño
analysts. This is dust blowing in your face and sand-dunes forming
underfoot, and dry lakes and rivers that don't flow, and green a
colour you only see in dreams. Have we made a mistake, have
the coordinates placed us on Mars? Back into the Tardis and back
beyond 30 000, and step out onto a green and pleasant land. Over

there a flock of ducks lands on a brimming full lake, and there the sound of a waterfall indicates a river in full flow; green grass is underfoot, and pushing through shrubs is your biggest problem. There is a distant rumour that far away, in the centre of this continent, is a small area of something called desert, biding its time, but it has no meaning here and now. And there, through the trees—hell, that's too big for a kangaroo, that's a kangaroo on steroids, 3 metres high, eating leaves in trees. And what's that over there? A rhinoceros, a hippopotamus, in Australia? No, but giant wombat doesn't seem to cover it—hang on, there's a giant wombat as well. And that glimpse, from the corner of the eye, of something in a tree—was it a leopard? Welcome to the roller-coaster world of late Pleistocene Australia.

There are two contradictory scenarios of what happened: mine, and that of Rhys Jones and Tim Flannery. In the latter, after humans arrive the vegetation is dramatically burned and the megafauna are killed off very quickly, human arrival being the defining moment. In my version, human arrival has no effect at all on fauna or flora.

This sounds simple, but the problem is highly complex, and this is why the debate about which is the correct version has been going on for around 150 years. The basic question is this: would the megafauna have become extinct if humans had never colonised Australia? If you think the answer is no, then you need to present hypotheses about the impact of humans on megafauna. There are two possible impacts, and they could have had an effect separately or together. The first is hunting, the second is habitat modification by the use of fire. A probable corollary of this belief is that the extinction must have happened within a very short time of human arrival. If you think the megafauna would have become extinct even if humans had never colonised the continent, then you need to present a plausible mechanism for the impact of the environment. Essentially, the only possible impact in this context is a change of climate. A corollary of this belief is that the extinctions must have occurred at a time of major climatic change.

In *both* cases you need to develop plausible mechanisms for the way the impacts caused the extinctions. You need to associate these with a theory about why this particular set of species became extinct, and you need to postulate the reason for the extinctions occurring at a particular time. On the face of it, the simplest of these requirements to fulfil is the last, and for this reason the timing of the extinctions has become a major intellectual battleground.

But the very first question is, what went extinct? Curiously, the way this question is framed influences the answer to the question about what caused the extinctions. The 'megafauna' went extinct. What is the 'megafauna'? They are the species that became extinct at the end of the Pleistocene period. Anything else about them? Anything suspicious? Well, they tend to be big, but that's about it. They are not all in the same taxonomic group, they are not the top twenty or thirty in terms of size (some that became extinct were smaller than some that didn't), although *within* particular taxonomic groups the larger ones were the ones that went. They weren't particularly odd (though some seem odd because they are extinct, in the way that say elephants, giraffes and warthogs would seem very odd if they were no longer living). They share nothing obvious in the way of likely food, habitat, or behaviour.

So they are not really a group. A better way of describing what happened at the end of the Pleistocene in Australia is simply to say that, whatever the mechanism involved, most species were not susceptible to it, but a few were, and they became extinct. The same pattern seems to have been true in all the other continents except Africa, where what we tend to think of as typical African fauna are the species that would have become extinct had Africa had some extinction mechanism operating. The problem has been that in thinking about the megafauna we have tended to refer to them as some kind of natural group. This leads to the following illogical conclusion. What happened to the Australian megafauna? They became extinct. What, all of them? Yes, every last one. Well, that certainly rules out climatic change as a mechanism, for if that was the cause you would think that a few of the megafauna at least could have found refuges.

Well, yes, you would, and many of them did—they were the ones (such as the large grey and red and hill kangaroos) that didn't become extinct! So no argument against climate here, but we still have to work out why some animal species became extinct at the end of the Pleistocene, just as we have to work out why dinosaurs became extinct much earlier, and why there have been numerous episodes in the history of life on earth when there were more extinctions than at other times.

So let us look at the simplest proposition first—humans caused extinctions by hunting. Why the megafauna? Well, the argument goes something like this. Everyone knows that it is very difficult to suggest hunting by hunter-gatherers as a mechanism for extinction. The problems are many. Hunters don't have an interest in rarity increasing value. As they successfully hunt a species and its numbers fall, it reaches a point where it is uneconomic to keep hunting them and people switch to alternative prey. The animals living in the relatively remote areas, or more rugged areas, survive and can begin to build up numbers again. In the meantime, the numbers of the new prey species begin to fall. If the original species recovers quickly enough people may switch back to it, or to a third species, and so on. In addition, as we have already seen, without modern weapons and modern transport and modern storage, and with relatively low population sizes, and with no ethos of killing for sport, the chances of a hunter-gatherer group causing extinction by hunting are remote. In Australia there is probably no example of it in the tens of thousands of years since the megafaunal extinctions. So, not hunting then.

What are the elements that make up the possible solution of this problem? The precise relationship between the timing of the extinctions, the timing of human arrival on the continent, and the timing and nature of climatic change has gradually become more focused as better evidence has been obtained. The position has gradually been refined to this: humans arrived on the continent not less than 50 000 years ago; in spite of a number of claims there is no evidence for a much earlier date, and on present understanding such a date seems very unlikely. Second, it now seems reasonably clear that the main period of aridity (considerably drier than now) in the last 50 000 years began, say, roughly 25 000 years ago and ended about 15 000 years ago. Conditions before that were considerably wetter than they are now.

So the battlelines are drawn quite clearly. If extinctions

occurred 50 000 years ago, immediately on human arrival, and during a period of very favourable environmental conditions, then there would be no doubt at all that humans had caused the extinctions, and the only debate would concern the mechanism. But if extinctions didn't occur until around 25 000 years ago, then human causation is out the window, and the prospect of any human contribution would be extremely shaky. And it only takes one good site to prove the case. It is why proponents of human causation have fought so hard to discredit every site that has suggested a young date for megafauna, and there have been many.

The main battleground has been an unprepossessing little swamp an hour or so north of Melbourne. Pleasingly, Lancefield Swamp was one of the first sites (in 1844) at which Australian megafauna were found, and it remains crucial in solving the mystery of their demise. In a nutshell, the megafauna are dated at less than 26 000 years at Lancefield, together with indications of environmental change, and circumstances that suggest a mechanism.[15] Every effort to shake the date from this site has failed, and I have heard that it has recently been confirmed. Also recently, another historic site, Cuddie Springs, appears to suggest late survival of the megafauna in northern New South Wales.[16] Other investigations on the Liverpool Plains actually suggest a very late survival of the megafauna, but this in a sense is icing on the cake. Lancefield tells us that any concept of simple human causation is defunct, and makes even human *involvement* look very unlikely.

Looking for large kill sites was originally a focus of much research in this field, and it was, ironically, originally thought that Lancefield might represent one. But it didn't, and the kill site is yet to be found—a circumstance, given the amount of research that has now been done, which suggests that it doesn't exist. In any case, the proposition was always dodgy, and relied more on inappropriate American and European analogies than any home-grown idea.

The idea was this. When French and British explorers landed on Kangaroo Island, unoccupied for several thousands of years, the kangaroos had no fear of humans and could readily be killed.

Promoters of overkill like Martin and Flannery have an image of the whole continent being one big Diprotodon Island, where naive large animals just stood still, from the Kimberleys to Tasmania, waiting to be killed by these strange new creatures called humans. The strange new creatures naturally preferred big animals—more meat—and were working their way from the top down and kept going until the surviving animal species became wary. This is a far-fetched fable, and a totally inappropriate use of analogy, as well as a misinterpretation of the way humans and animals behave. It is also a misunderstanding of the difference between an event lasting a few days and one lasting thousands of years, and between events on islands and events on continents. Alternatively, we have the dopey Diprotodon suggestion. In this one the idea is that the megafauna were the slow and stupid members of the Australian fauna, very easy to kill, and selectively sought after both for that reason and because they provided more meat. Eventually the easy pickings just ran out.

But the megafauna are just bigger, stronger, possibly faster, and tougher versions of the species that survived. It is hard to see the giant kangaroos, for example, as easy pickings, whatever your African analogy is for the Diprotodons. (Rhinoceros? Not many people have made a living eating rhinoceros!) There is no evidence that Australian animals are inferior to other animals in intelligence or adaptation to the environment. Even if Diprotodons were slower or more stupid than other species (and remember there is no evidence for this), hunter-gatherers still don't keep hunting increasingly rare species while more abundant ones are staring them in the face. If hunting 'overkill' was the mechanism, all the megafaunal species would have survived to the present day—possibly in relatively small numbers, in places that were hard to get at, but nevertheless surviving. If hunting and 'overkill' was the mechanism, there would have been a continuing progression of more and more species becoming extinct, as each in turn died in its Alamo.

Then what about the environment as the mechanism for extinctions? Drought, whether the last one, the current one, or the next one, is on the mind of most Australians, part of the uniqueness of the Australian continent and the Australian character. So when the nightly news in summer is as likely as not to show starving sheep and cattle, it is the logical mechanism to look for in explaining megafaunal extinction by climatic change. But there are several difficulties, and a theory has to explain why these particular species were affected and not others, why there were no ecological refuges where species could shelter, why the geological record doesn't show a matching catastrophic vegetation change, and why the extinctions occurred when they did and not, say, 100 000 years earlier. The last question is dealt with separately, to maintain suspense, but let us look at the other questions, which are related.

The difficulty in the past has been that people have tended to think of the mechanism as being related to food supplies, and if this were the case it is impossible to see why there should be selection against larger animals, and why an event of such catastrophic magnitude as to wipe out vegetation would not be as visible as the evidence for a meteor hitting the earth. Nor is it possible to see why there wouldn't have been some areas of refuge where vegetation survived (hard to see a drought affecting the *whole* of Australia simultaneously). While people have wondered about water supplies, it is impossible to believe that there would be no water available anywhere (no Murray River? no springs?), though it has been demonstrated that many large lakes had dried.

The mechanism, I reckon, is this, and it does answer all the questions. Small animals can find much of their water needs without drinking, making use of water in food, and dew, and the puddles formed after rain. Such animals living in dry environments also have behavioural and physiological mechanisms for making the best use of water they can find and conserving water in their bodies. Some larger animals, including the desert kangaroos, also have these attributes. Now what if the megafauna were the species that didn't? What if the largest members of each taxonomic group were the ones that couldn't survive without drinking each day (or at some regular interval)?

Well, most of the time you would get by just fine. You would come in to a lake to drink each night, and spread out again each day. When food began getting scarce near your favourite billabong you could move on, drinking from creeks or small lakes until you reached another good water source with plentiful food. But what would happen in periods when all the ephemeral water supplies had dried up, and permanent water supplies from the largest rivers, or the occasional springs, were long distances apart?

Imagine a mob of Diprotodon near a permanent spring-fed waterhole. The grass has been eaten down to just a few stalks, and even woody vegetation in unpalatable shrubs has been chewed back. Every so often the leader of the mob sniffs the air, and we see the mob head off in some direction towards where, on far distant hills, there appears to be some thick vegetation, but within a day they are back, desperately drinking from the waterhole, clearly thirsty and distressed. They repeat this procedure every few days, but each time the interval before their return becomes less, and their weakness becomes apparent. Even the woody shrubs are gone now, and the thin dry stalks provide little for such large animals. Soon one lies down near the edge of the water and fails to get up again. In the days to come others fail to get up in the morning. The last one dies helplessly, actually in the water taking a last drink, pitifully thin—it may be that the last drink, of a warm soup of blue-green algae, or some nasty bacterium, is what administers the *coup de grâce*.

Now look at the overall pattern. Conjure up the giant bullseye mentioned in the previous chapter, with red desert in the middle. When the climate changed so radically in the late Pleistocene, the dry heart of the continent expanded outwards, and all the circles of different shades of green surrounding it expanded outwards too.[17] Even at its greatest extent, though, there was still a broken fringe of green on the coast and ranges, surrounding the huge central desert. The green formed a series of refuges for all kinds of species, and here and there among the trees could be seen the sparkle of water. But although it looked as if here was a fleet of Noah's arks, ready to carry the genetic inheritance of Pleistocene Australia on for at least another 25 millennia, those arks were not equipped properly for large animals.

If you have to drink free water, and if the next free water is too far away for you to reach in the time before you need to have a drink, you will be tied to that water as surely as a lawn-mowing goat is tethered with a chain. On that tether you will eat out all the available food and then die. The water will still be there, and you may in fact die in it. There will still be unchanged vegetation all around. There will still be species of small animals surviving quite happily. What is worse is that the effect would come upon the megafauna and kill them before they could take evasive action. You can't move and stay ahead of the expanding desert, because the effects of thinning out the water supplies locks you in place where you are, before the desert arrives or is even sensed as coming. When the desert again retreats, the surviving species can expand their ranges once again to occupy the plains, but the megafauna are not there to move back.

What about Aboriginal firesticks, you might still ask. Since there's no doubt that fire does have a major impact in Australia, what if the megafauna were the species susceptible to the modification of the environment by fire? It's a plausible theory, but on closer examination it makes little sense. Even if we didn't already know that human use of fire has had precious little impact on the Australian environment, a particular role for fire in relation to megafauna doesn't make sense.

The thing that everyone agrees on is that after a fire there is a period when grass grows well, having regenerated first, and that this is advantageous to kangaroos and large wallabies, as we've seen. It would equally have been an advantage to the megafauna. Bad for small animals, because their shelter is destroyed by fire, and they often don't eat grass, but good for large herbivores. Even if you were to argue that all the megafauna were browsers (and there is no evidence that they were) and all the surviving species grazers (which they are not), regrowth of small shrubs and leaves on trees are equally well promoted after a fire. If creating grasslands in Australia was good for cows and sheep, such effects would have been just as good for the Diprotodons. And finally, the time of arrival doesn't make sense. Why would the use of fire have no effect for some 20 000 years, and then have an effect?

It all sounds very convincing, you might say, but can it be true? You talk as if this has all just happened once, but there have been many, many occasions when the desert has expanded out and back, and all those other times were just like the most recent one. The extinctions didn't occur during the earlier expansions and the only difference is that humans were here for the last one.

Yes and no. The idea that all the climatic fluctuations of the Pleistocene were much of a muchness was a function of the generalities of the initial stages of geomorphological investigation in Australia. And it seemed convincing—you can't argue for a climatic mechanism if the climatic change 25 000 years ago is no different to anything that went before. But I put up the proposition anyway, rather in the spirit of astronomers postulating the existence of Pluto from its effects on Neptune, but unable to see it. Powerful telescopes vindicated the existence of Pluto, and some new powerful techniques have vindicated the climatic theory. Two new players, Gerald Nanson and Paul Hesse, have come in with new analyses and demonstrated that the late Pleistocene was much more severe than what had come before.[18] Paul Hesse in particular has studied the dust, blowing out to sea for hundreds of thousands of years (from a continent that can't afford to lose a cupful) and settling down through the water to the seabed between Australia and New Zealand, disturbed occasionally by the flap of a fish's tail. Stronger winds and greater aridity at the end of the Pleistocene than ever before, says Hesse, a conclusion matched by Gerald Nanson's analysis of river sediments and his conclusion that this was the driest period of the Pleistocene.

The 'Drunkard's Walk' is a metaphor from chaos theory and evolutionary theory.[19] The drunk weaves down the pavement from left to right, but on the left is a brick wall, so he can't move far that way; on the right is the gutter. Backwards and forwards he goes, narrowly avoiding disaster each time until one quiver of the muscles, or a little puff of wind when he is on the kerb, and he falls into the gutter and the walk is over.

The Drunkard's Walk, I believe, is the explanation for extinctions in Australia. It means that the difference between surviving and not surviving is a small quantitative increment, not susceptible to analysis by archaeological means—there is no smoking gun. In the case of the megafauna there are two related drunkard's walks, the baby and the bullseye. The baby is El Niño, or previous similar phenomena, and the movements of the southern oscillation index are a clear example of a drunkard's walk. Up and down goes the squiggle on the graph, the result above-average rain, then below-average rain. Every few years, just a bit more difference and below-average rains become a drought year. Up and down, up and down, a few years this way, a few years that way, then, randomly, five years that way, ten years this way, fifteen years that way. Even in a land with artificial dams and bores, farmers and their stock have trouble with droughts lasting that long. Water dries up entirely, and it is too far to walk to the next one, or it partially dries, and the last muddy puddle is a toxic soup of blue-green algae or virulent bacteria. Sheep and cattle die in the mud, too weak or sick to crawl out, or die on the edge from poison or disease or fear of braving the mud for a last drink. All around the once plentiful grass has given way to bare earth, as the concentrated animals, tethered to the water by the need to drink, desperately try to toss up between the need to drink and the need to eat. But lack of water will kill you faster, and lack of food will gradually be compensated for by a shrinking gut and wasting muscles, until the point of no return is reached.

The bullseye is the theory of Richard Gould, who noticed that because of Australia's approximately round shape, and its identity as the smallest continent/largest island, and its mid-latitude position, the climatic and therefore vegetation zones were arranged in roughly concentric circles extending out from a central desert to wet forests around the edge. It looked like a target, the desert being the bullseye.[20]

It became plain from the work of palaeoclimatologists[21] that while the target had maintained its shape over a very long time, the bullseye had been both smaller and larger in the past, and as a consequence the concentric zones had also moved in and out. Most of the time they just oscillate around an average point much like conditions today, but at times of maximum aridity the wet fringes are greatly reduced in extent, and it would only take one drunken lurch to have the species surviving in fringing enclaves pushed over the edge to extinction.

The proposition, then, is that climate has been fluctuating in Australia for millions of years, and species have changed their distribution patterns and abundance in order to adapt to those changes. But Australia's design is such that it has always been a close call, the megafauna teetering on the edge of the table, only to roll back towards the middle at the eleventh hour. But just a slightly bigger push when the drunkard spins the bottle, and it spirals towards the edge and falls over to smash on the floor. The end of the Pleistocene is that period of the smash then, the environmental equivalent of the stock market crash of 1929. Just a little bit drier on the margins, the desert just a bit bigger, just a few less active rivers and waterholes, and there are massive impacts on a few species that had survived hundreds of thousands of years of smaller fluctuations previously. And the presence of humans is of no more significance than that there was an audience for the losses. This overall pattern, incidentally, with its gradual and cyclic changes of climate over hundreds of thousands of years, puts paid, if this even needs saying, to any notion that humans cause climatic change in Australia.

And what about other parts of the world? Close to home, it might be all right to suggest something like this for the Australian continent, but surely no one is going to suggest that those large green islands at each end, Tasmania and New Guinea, could have been part of this? How on earth could you use drought to explain the loss of species in these areas or in Indonesia? Well, everyone knows that it can get pretty dry in Tasmania, and there are ancient sand-dunes to bear witness to the same sort of climatic stress as on the mainland. But New Guinea?[22] In 1998 CARE Australia fed starving villagers in New Guinea, where a year-long drought withered the crops and fires burnt through dry forest. In Indonesia thousands of square kilometres of tinder-dry forest burnt following the worst drought on record. And this was in a considerably less dry period than was the case in the late Pleistocene. This was a particularly bad El Niño. But any megafauna would have struggled in that year, and would have become as surely extinct as any in

Australia had the situation been worse and longer lasting. The Greenhouse effect is likely to duplicate that terrible time.

What about extinctions elsewhere? It is possible that humans caused extinctions in New Zealand,[23] though I don't think environmental change has been ruled out. It is possible that humans caused extinctions in North America, though again I don't believe that climatic change is ruled out as a cause there. In both cases the environment, the animals concerned, and the culture, technology and society of the humans was different to Australia, and can tell us nothing about our continent. Certainly nothing seems to have caused extinctions in Africa. I don't believe this has anything to do with evolving together, but more to do with the fact that the African continent, astride the tropics, is the reverse of Australia. In Africa the central core is wet, the outer parts dry, and the result may well be that there are always potential refuges.

In the trees on the hill overlooking my house are the graves of three lambs that died at birth. All around are the lambs that lived, and sometimes when I walk through the trees I fancy I can see the ghosts of those three lambs running with the others. But two other lambs were taken by a fox, and I never found a body to bury. Those lambs I can't picture as ghosts. Somewhere down in the broad valley to the south, the bones of Diprotodon lie buried in sediments washed down from the hills. Above them the young kangaroos play, and it is not hard to imagine the young Diprotodon conquering their history and joining them, their ghosts seen passing by a billabong. But if humans had killed them there would be no ghosts, no shadowy large figures in the trees; the ecosystem we see now would have been arbitrarily altered.

Would it matter if the Australian ecosystem of today had been arbitrarily altered? Well, surprisingly perhaps, not much from the point of view of the environment itself. Of course the loss of any species has ramifications of a greater or lesser extent. In the case of the megafauna, for example, the loss of such large herbivores may well have affected the ability of Thylacoleo to find prey. Its demise, in turn, may well have meant that there were fewer carcasses to be scavenged by Tasmanian devils. But because these animals were so large, their use of the environment was very coarse-grained. Take some large herbivores out of a system, and all that happens is that other large herbivores increase in number, or numbers of different kinds of invertebrates increase to take advantage of the available vegetation. Balance is maintained in a new set of relationships, and no matter how fast the extinction process there would have been no recognisable change in the vegetation. Whatever slight adjustments there might have been would have been greatly outweighed by the effects of climatic change.

So, does it matter? Well, as in many areas of political life, and life generally, perception is more important than substance. If extinctions were caused by humans, then the Australian environment we see today would be an artificial construct, and artificial constructs have no independent reality. A landscape dramatically altered once can be altered again and again. A similar argument can be seen in the idea that if a forest has been logged it is no longer 'natural' and therefore not worth saving—the woodchips are on their way before the echoes of the word 'regrowth' have faded. In the mountains, if cattle have grazed an area then they should be allowed to keep on grazing, no matter the massive damage.

Rhys Jones, as always, neatly encapsulates the proposition: 'what do we want to conserve, the environment as it was in 1788, or do we yearn for an environment without man, as it might have been 30 000 or more years ago?' Rhys goes on to use this as an argument for control burning, and I shall return to this. But it is the same idea here: if humans caused extinctions, then we needn't look at the present Australian environment as a natural system to be conserved. But humans didn't cause extinctions, and the environment in 1788AD or 1788BC, or 17 880BP, or 37 880BP is the result of the environmental history to that point, and we need to maintain Australia in that evolutionary state. What we have is the way it was meant to be and there are no excuses.

Furthermore, if climatic change can get rid of such a chunk of Australia's mammal fauna in the past, then it can do so again in the future, and this time we would be starting from a much lower base. The 1998 Kyoto conference reporting on Greenhouse effects seems to have largely focused on sea-level change, natural disaster increase, and the impact on agriculture. But if, as seems likely, Greenhouse can interact with El Niño to produce much more drastic and longer-lasting El Niño events, then we are likely to be in for other dramatic extinction events. And while you can remove large herbivores without very dramatic effects in Australia, the next round will be smaller herbivores and consequently carnivores, and the impact on biodiversity will be much greater and the chance of maintaining our fragile environment much less.

In a famous phrase, which caught the world's imagination, chaos theory was summarised in the happy thought that a butterfly flapping its wings in China could ultimately cause a tornado in America. In many ways the world is as tough as old boots. All kinds of changes have occurred over billions of years, but it has just kept rolling along, that old man Earth. The strength of the system is its complexity, and its capacity for adjustment and repair. Aborigines recognised that environmental strength and sustainability come from biodiversity, and maintained that biodiversity by recognising that every element in an ecosystem is important. There is no short-term profit in biodiversity, just the reverse. But those pushing commercialisation and the monetary value of wildlife are like those who pushed potatoes onto the Irish 200 years ago. In the short term, a monoculture is profitable and feeds the peasants, but in the long term there is always a blight waiting for those who think the environment should be conquered for the profit of a few.

You can push any system too far and it loses its elasticity, its capacity to bounce back. The weakness of the system is that it is vulnerable to particular types of attack, but also to catastrophic events, and, as the well-known phrase goes, 'extinction is forever'.

The extinction of the dinosaurs may have been a catastrophic event almost beyond human imagination,[24] a small dot of light in the sky becoming larger and larger until it almost filled the sky and then struck the earth with a massive explosion. The extinction of the megafauna occurred more with a whimper than a bang. Here is chaos theory at its most subtle. A fish flaps its fins off the coast of Chile and a Diprotodon pauses and sniffs the air and wags its tail. The red centre itself flickers in and out, like the tongue of a Mountain Devil flicking up ants, and picks up the megafauna.

Okay, so Aborigines didn't change the environment by the use of fire, and they didn't cause the megafaunal extinctions. I guess that makes them the first true conservationists. Well, yes, but on the other hand, no.

7

'Most enlightened conservationists'

Traditional Aboriginal attitudes to land are not very different to those of the most enlightened conservationists. But they are based on a far deeper and far older association with it than any we can claim.

—JUDITH WRIGHT, 1983[1]

Ironically, in view of the attitudes of the Liberal and National parties towards them, Aborigines were probably the world's greatest conservatives. It is also a strange twist that sees both the conservation movement, and most Aboriginal people, on the left of the political divide. Aboriginal approaches to the land seem to me to differ from those of both conservationists and political conservatives.

It is yet another irony that has seen Aborigines damned for their conservatism by political conservatives—why, they hadn't even got round to inventing the wheel, Hugh Morgan sneered. Nor had they invented that ultimate vehicle of change, agriculture. A view from very early times to the present day has seen Aborigines as being locked in a time warp, existing in stasis for 50 000 years in this isolated corner of the world while the rest of the world advanced around them.

It was difficult for those on the Left sympathetic to the Aboriginal cause to argue against this proposition, given the technological and economic state of Australia 200 years ago. Many people would add state of society to the evidence for lack of progress, but if 200 years of Aboriginal studies has achieved anything it has demonstrated the enormous complexity of Aboriginal society. All modern societies are advanced societies; Aboriginal society is more advanced than most. It could be argued that all peoples are faced with a choice about whether they put their energies into developing their economy/technology or their society. Aboriginal people were not terribly interested in the former as long as it was sufficient to keep them fed. They were terribly interested in souls and arts and kinfolk.

Nevertheless, in the Olympic events for top societies, the brute force of advanced technology and booming economy beats wishy-washy artistic, social and religious values every time. Indeed societies which have made that choice in various parts of the world (Tibet, Wales, Ireland, Cambodia and Brazil, among others, as well as Aboriginal Australia) have usually been stamped on by their neighbours or by colonists. The Athens/Sparta division is an ancient one.

Faced with the dilemma of supporting a people who hadn't played the growth game, Aboriginal supporters were forced to argue that Aborigines hadn't pushed the progress button because they were conservationists. In conscious or unconscious recognition that Australia was a fragile continent, the Aboriginal religious system had developed in order to conserve the environment. They believed that animals and plants should be protected, and this was an example to us all. Unfortunately, while I think that the nature of Aboriginal society conserved the Australian environment for 50 000 years or more, I don't believe Aborigines were conservationists in the current Western sense of the term. The contradiction between these two views is explored in this chapter.

If Aboriginal people had been acting as conservationists, there are four different ways in which this could have happened. The first way would be totally up-front, conscious, all-out conservation behaviour, in which cause and effect and long-term consequences were all assessed and activities with potential negative consequences adjusted accordingly. Such activities might include, for example, deliberately refraining from killing female animals (or those with young) in recognition of their role in continuing the species; deliberately ceasing to hunt or gather when it became obvious that a species was becoming rare, in recognition of the fact that rarity is next to extinction; avoiding any modification of the environment that would cause harm to particular species. Equivalent behaviour in the modern conservation movement would be the protesters who try to stop duck-shooting, and those who try to prevent the clearing of koala habitat.

The second way would involve positive conservation activities. Such things as planting or replanting the vegetative material of plants, or modifying the environmental conditions to favour a particular species, would come under this heading. The classic modern conservation approach under this heading is the Landcare program with its tree-planting projects; another is the captive breeding of species in enclosures that keep out foxes and cats.

Under a third heading would come actions based on religious and other philosophical beliefs which unconsciously have positive environmental outcomes. It is hard to think of comparable Western practices, but vegetarianism and tree worship can have environmental outcomes, as can the preservation of country churchyards.

The final way involves actions taken for economic reasons that have conservation outcomes, whether consciously or unconsciously. Failing to hunt the last (or preferably second last!) wallaby of a particular species may be the result of the enormous effort that would be needed to find and kill animals of such rarity. It would make no sense to do so, and hunting would switch to more abundant prey. This means that rare animals can have the opportunity to rebuild numbers, but no decision has been made with the intention of enabling them to do so. The best Western example is fishing, where commercial fisherman switch from one species to another when the first becomes too rare. It is not from a desire to save a particular tuna species, but simply that fishermen can no longer afford to target them and make a living.

When people from the traditional environmental movement

refer to Aborigines as being conservationists they are thinking of actions in the first two categories. They believe that Aborigines routinely behaved in these ways and that ethnographic evidence supports this belief. They also believe that Aborigines behaved in this way because of a mystical connection with land and Nature, a belief shared by many Aboriginal people. The conservatives from the new hard Right of the environmental movement, on the other hand, believe that Aborigines behaved in a totally self-interested way, and that whatever conservation outcomes there might have been were the accidental and incidental outcomes of economic rationalism in hunting.

The third conservation mechanism has been little examined, the few examples proposed probably representing 'economic conservation'. But this category probably holds the key to living on this continent for 50 000 years. It is subtle, and not very glamorous, but this is almost invariably the case with non-populist environmental answers.

And so to the evidence.

Dick Kimber has been a major proponent of the idea of Aborigines as conservationists ('there was also a genuine notion of conservation. Animals were clearly recognised as "belonging" to particular habitats'), and he has put forward two examples from central Australia to support his belief, both of which would come under the first option discussed above. On one hunting trip, immediately after an emu had been shot, a kangaroo was seen and not shot. The people Dick was with said, 'Let him go—we got enough. That his proper country. We get him another time.' On another trip, after *four* kangaroos were shot, the shooting stopped, and the explanation offered was similar: 'He your countryman. We have enough. This his country too. Let him go. We can get him any time.'[2]

But the most that could be said about these examples is that they illustrate the proposition that Aborigines probably didn't normally hunt for sport. In both cases, however, it is clear that the motives are economic and that the conservation gloss is a rationalisation after the event. Kimber himself notes that without a truck there was a limit to what a hunter could carry, and even with a truck, excess meat would quickly go bad. You have to ask yourself, if the emu had not been shot previously, and the other four kangaroos obtained, would the two kangaroos concerned have been allowed to escape? The answer I think is obvious, 'country' or no country—they escaped because the hunters 'had enough', not because of any ethical concern.

Kimber also notes another practice in the desert that was the reverse of conservation. When hunting, people had an 'apparent preference for doe kangaroos instead of bucks. The latter were sometimes referred to as the "breeders" . . . it is considered preferable to shoot a doe rather than a buck.' Kimber thinks this is an aspect of conservation behaviour, but his comment that 'it does not necessarily agree with European knowledge or views' is a giant understatement.

And now to the second category of possible conservation activity, the deliberate propagation of plants. This has received the most attention because it relates to the proposition that Aborigines were well on the way to developing farming, a development nipped in the bud by the arrival of Europeans. Dick Kimber has paid considerable attention to this, but many of his examples of the movement and planting of seeds are such singular events, and are so clearly the result of European influence—for example, in 'non-traditional' circumstances—that they tell us little.

In a later discussion, Kimber links together mythology with an account by an old man called Walter Smith.[3] The mythology is to do with the transport and spreading of seeds in the Dreaming, but as Kimber notes, the 'descriptions explain the natural distribution of certain plant-foods'. That is, rather than being a description of human activity brought down from the past, they represent an attempt to explain the results of a natural process.

Walter Smith's account of historical activity also rings strangely: 'They chuck a bit there [at a favourable locality]. Not much, you know. Wouldn't be a handful . . . one seed there, one seed there . . . they chuck a little bit of dirt on.' Now this would be fine, in a sense, if the seeds concerned were of, say, fruit trees. They are, however, *grass* seeds, and even if the story were true, the impact of such activity would be minimal. In addition, it is the recollection of just one old man. There is no knowing how much of the description is based on knowledge of European practice, how much on the mythology, and how much is Smith's own attempt to explain a naturally disjunct plant distribution.

Rhys Jones has also had an interest in plant propagation as a precursor to agriculture, but it is the conservation implications that interest us here. Here is his description of how the parsnip yam, *Dioscorea transversa*, is harvested in Arnhem Land: 'when the bottom of the tuber has been reached, it is snapped by lifting upwards, and is removed, leaving the top of the tuber still attached to the tendril intact in the ground. When asked why they did this, people answered that soon the yam would grow again and they would return and dig out some more.'[4] There are similar descriptions from Cape York, where people also maintain that the technique 'develops a multiple-ended tuber, a *thampu paapay* ("mother yam") which is especially succulent'.[5]

This latter description suggests a motive that has more to do with taste than conservation, but there are in any case difficulties with the whole process. Jones describes the harvesting technique as digging a hole which is 'about 10cm away from where the tendril enters the ground and is obliquely angled so as to meet the yam tuber about 15cm underground'. This sounds likely to be just a matter of practicality—the withered tendril can't be followed directly down because it is too fragile, and therefore the aim is to hit the tuber itself. Once this approach is followed, the natural consequence is to break off the tuber you can reach, rather than bothering to follow the whole lot back upwards. It would be interesting to know what the practice is in times of food shortage.

The reasons attributed to such exercises may amount to little more than an attempt to give added meaning to a process that is purely mechanical. Confirmation of this comes from Jones's observation that 'in actual fact, despite their stated intentions, people quite often did not return to the same place'. Further confirmation comes from his description of the hole being 'left open, but in that environment it would soon fill with friable soil and vegetable matter'. It doesn't take much to push soil back into a hole if your intention really is to let a plant regenerate.

The only other proposed examples of deliberate propagation seem to relate to fruit trees. It has often been noted that bringing fruits back to camp results in trees growing from discarded seeds, but Jones takes this further: 'man's effect is not merely a matter of dispersal: the seeds are deposited in specially favourable environments, the midden soils with their compact heaps of decaying organic material and lime from shells.' None of this amounts to more than the accidental dispersal of seeds, in the same way that most animal species help to disperse seeds of all kinds of plant species. Hynes and Chase suggest that in some cases seedlings in camps 'were protected by ground clearance and by erecting small barriers to prevent children inadvertently crushing them'. But it is difficult to see this happening in 'traditional' non-permanent camps, and it is hard to be sure that some of this activity wasn't the result of the observation of European activities in gardens.

In a world increasingly governed by economic rationalists and their disciples among the conservative politicians, the idea that the economy is not just the main thing but the only important thing in human affairs has taken over political debate. In the conservative mindset, it makes perfect sense that species can be saved by exploiting them, and the environment only needs to be saved to the extent that it is valuable to the economy. In this view, then, the idea that Aborigines only conserve things accidentally, as a consequence of economic and technological considerations, makes perfect sense, and serves to counter the arguments of the 'rabid greenies' (demonised by conservatives), who use the Aboriginal attitude to the environment as an example to Western society. Certainly it is true that much apparent conservation behaviour and outcomes are the result of the imperatives of hunting and gathering, but the situation is rarely simple, as the following examples illustrate.

Optimal foraging theory has been studied in relation to the real hunting and gathering behaviour of the Alyawarra people of central Australia.[6] The theory is complex and subtle, but it can be summarised in a simple way. Food items are valuable to people because they provide different amounts and proportions of protein, fat, carbohydrate and other essential elements of the diet needed to keep people alive. But food comes at a cost: the cost of searching for it, handling it, and processing it. Other things being equal (which they are not, but bear with me), people will preferentially obtain food that has the best nutritional return for the lowest cost. Furthermore, if the ratio between cost and return changes (such as when so much of a particular food has been collected that it becomes rare and therefore harder to find), people switch to a different food.

One time the Alyawarra switched from Ipomoea to Vigna. 'We suspect foragers had exhausted Ipomoea . . . or at least reduced its abundance to the point that Vigna in the adjacent patch became the optimal choice.'[7] That is, as a more favoured species was exploited more and more it became harder and harder to find; as the cost of searching for it rose it reached a point at which the return simply wasn't worth the effort, and a less favoured species began to be collected. The abundance of the first species would then have a chance to build up again, while the second one fell. Obviously things would normally be more complex than this, with a large range of different species, and varying degrees of abundance and preference. The more complex the diet, the more chance there is that the system can result in all species within it being maintained by this mechanism.

The rich environment of Arnhem Land has a system that appears to be the reverse of optimal foraging: 'a strong contrast between the broad spectrum of shellfish available to the Anbarra and the narrow target [that] reflects consciously expressed choice.' After a period of heavy rain that destroyed the beds of the favourite species 'other species of shellfish available in enormous numbers . . . were as far as I know not eaten'. Shellfish species occupied 'a special place in Anbarra culture which is not due solely to their nutritional content'. Anbarra don't switch when things become rare but will keep after them even when others are available. Such behaviour could lead to species extinction.[8]

On the face of it these two studies are contradictory, and provide a salutary lesson, first in the risks in using ethnography to reach conclusions about environmental matters, and second in the risks of thinking that Aboriginal Australia is uniform.[9]

Part of the problem, of course, is that man does not live by bread alone. Food is valued at one level not just for its nutritional value but for its taste, palatability, texture and so on. At yet another level food can have added symbolic values associated with such things as status, image, gender, social relationships. Finally, there can be religious significance given to foods or the places they are obtained from. The effect of all this added value is that people will spend much more time, put in much more effort and pay more for certain foods than could be predicted from their nutritional value alone, and they will go on trying to get them even when easy alternatives are staring them in the face.

But they won't go on trying to get them when starvation is the alternative, and this is the other way in which these two apparently contradictory studies can be reconciled. There are abundant resources in the north: 'asked about the resources of any place, particularly their own country, Aborigines of the north invariably answer in the creole expression "can't finish him up". The main reason that this expression is true is that they maintain their populations at low levels that allow them to continue to depend substantially on high yield/low probability [animal] foods.'[10]

Let us take a risk and simplify what is a complex matter. Whereas in rich environments Aboriginal groups could keep their populations low enough in relation to resources to have the luxury of concentrating on favoured resources, those in poor environments (notably central Australia) are simply unable to do so while maintaining viable population sizes. In poor environments, then, conservation occurs because people switch to new foods when numbers of favoured foods decline. In rich environments conservation occurs because human numbers are so low relative to the abundant resources that they can't cause extinction even when they don't switch. 'Can't finish him up' becomes a description of conservation as much as of country.

So, economic rationalists rule. Aborigines didn't cause extinction of species in Australia, but not because of any mystical wishy-washy liberal humanist greenie view of the environment, just a matter of cold, hard economic reality. If there had been a dollar, or a meal, in the last pair of Paradise Parrots 3000 years ago, they would have been shot ducks.

And yet, and yet, there must be more to it than that, mustn't there? The close links between Aboriginal people and the land, totemism, the Dreaming, sacred sites—surely there must be something in all that which equates to conservation. Well, people have thought so. Alan Newsome analysed the distribution of red kangaroos in central Australia, and the distribution of totemic sites for the species, and observed that the sites represented areas of the most favourable habitat for the species. Hunting was forbidden near the sites, and therefore, Alan concluded, 'red kangaroo were protected near their best habitat'.[11] This example is often quoted but it is virtually unique. It looks, on the face of it, very much like the modern conservation practice of establishing parks and reserves. David Bennett, for example, says: 'Newsome's analysis . . . and Strehlow's observations that the totemic sites form game reserves, suggests that the Arunta built a conservation principle into their moral system, or more forcibly built their moral system in part on a conservation principle.'[12]

It all seems a bit too good to be true, and it probably is. Totemic sites are almost always distinct features of the landscape— waterholes, rocks, caves, hills, rivers and so on. Some features of this kind will correspond to features that are important to some animal species. Waterholes, for example, are of prime importance to many species, and protecting both them and the surrounding areas will have benefits to both animal species and humans, particularly in dry areas. But while there is no doubt that waterholes are crucially important in the centre, this was certainly not true of the well-watered parts of Australia. People in such areas may have had mythology associated with water, but there is no doubt that creeks and lakes and springs also featured as major hunting, fishing and gathering areas. Where water is limited it becomes a major factor; where it is not, food-gathering has equal importance.

Totemic sites are particularistic, not classificatory. While individual waterholes or rocky hills may be protected, not *all* such sites in an area will be protected. And in any case, sites are sometimes protected only against particular clan or other groups, leaving other people in the area free to hunt in them. Because it is sites that are being protected rather than the species themselves, a species that, say, sheltered in a rocky outcrop but fed on the plains could be hunted to extinction even without hunting in the outcrop. Sites are rarely concerned with environmental zones, so unless a species finds all of its needs within a natural feature (say a cave) it is not being effectively protected.

I know of only one example where it is said that a particular environmental zone is protected, and this is the relict vine thickets in northern Australia. Rhys Jones has suggested that such stands are deliberately protected from fire, and if this was the case, species that live in that habitat would also be protected.[13] But it is by no means certain that this is actual protection, and not just a recognition after the fact that such areas are often in fire 'shadows' and do not in any case readily burn. It is reminiscent of the claim that clapping hands will successfully keep elephants away—you don't see any elephants, do you?

So it looks as if totemic sites neither serve to effectively and totally protect species accidentally, nor were they intended to do so. But irrespective of the motives, it is outcomes that are important, and such areas did provide some level of protection, a level that may have been critical for some species. They were treated as inviolate areas of land, totally protected by the full sanction of law, even though they could have been important economically. Our national parks, wildlife refuges and wilderness areas have tended to be areas of no use for anything else. Furthermore, if they do become useful for something else, like logging or mining, then that takes priority, as the recent disgraceful decision on uranium mining demonstrates. They needed to be treated as our sacred sites, inviolate areas of land, totally protected by the full sanction of law.

Is there nothing else in the Aboriginal world view that promotes conservation? The Yarralin people of the Northern Territory 'see the nurturance of all life as the ultimate achievement of the Dreaming . . . The cosmos can be seen as a closed, self-regulatory system which seeks a steady state in which all life is maintained at optimum levels of productivity.'[14]

Another observer, David Bennett, says Aborigines 'feel that the order of their physical universe is interconnected with and incorporated into their social order. To maintain certain aspects of their social order it is necessary to maintain related aspects of their physical environment. This means that if they were to allow, say, leeches to become extinct by omitting the proper rites, a social consequence would eventuate.' (I feel the same way, but it is not a common world view in Western society, though chaos theory has converged on it.) He points out that non-human species are regarded 'as having equal rights to the resources of the country and as being necessary to the well-being of the country as well as to human well-being' and that there 'is a recognition of a unity and mutual interdependence that establishes the moral relevance of non-humans to humans and vice-versa'. This differs sharply from the 'Judaeo-Christian/Stoic traditions in which humans and non-humans were separately created and the supposition from this separate creation that humans differed from non-humans in form and essence'.[15] Aborigines postulated the existence of DNA long before Watson and Crick, striving for precedence of discovery with Rosemary Franklin in their Cambridge laboratory.

But the problem with all this is that it doesn't translate into conservation action—or belief, for that matter. For example, while Aborigines may have escaped the poisoned chalice of being given 'dominion' over the rest of the universe, they also have a belief in a kind of a place for everything and everything in its place, which differs little from dominion or the ladder of perfection. Aborigines see animals (and presumably plants) as being 'the foundation of their subsistence'.[16] That is, it is a human role to kill animals for food, and an animal's role (rather like the schmoo in Douglas Adams's *Restaurant at the End of the Universe*) to be eaten. The subtle difference, it seems to me, is that while some people (and not just in the West) believe that it is laid down in the Bible and similar books, the Constitution of the Human Race, that every human being has the right not only to carry arms but to use them, if they so choose, to kill every last Tiger, Passenger Pigeon, Dodo, or Toolache Wallaby, Aborigines don't believe this, and would not choose to behave like this. On the other hand, there is nothing to prevent an Aboriginal person, if hungry, killing and eating the last Toolache wallaby.

Finally, although Aboriginal people believe that the universe should be unchanging, this seems to me more a belief that the universe will maintain itself than an injunction to roll up the sleeves and make sure it does. Sure there are ceremonies and increase rites—scattering bits of shell in the sea of Arnhem Land to regenerate the shell bed, for example, or beating rocks where the specks of dust become kangaroos after rain—but these are self-fulfilling prophecies in nature, not actions with direct practical outcomes.

It is easy to be cynical about this, but ultimately it is, I think, religious belief that ensured that the Australian environment was conserved. Yes, there was a belief that the universe should maintain itself, an unchanging land to support its unchanging people, world without end. While there was no injunction to get out there and rake and prune like so many gardeners in the ancient stately home of Australia, there was a belief that you didn't do things to damage it. Aborigines would not, for example, have drained a mosquito-ridden swamp (or sprayed tonnes of chemicals to kill mosquitoes, as the NSW government is in the process of doing), or set out to exterminate dingos. They travelled lightly through the land, as though with an injunction rather like the Hippocratic Oath: 'First, do no harm'. And this is a very great deal indeed. If you believe the universe has always been the same, and always will be the same, you will do much less damage than if you believe that land is there to be developed.

Aborigines would also not have set out to burn the country in order to develop it. The firestick farming theory seems to have prevented people seeing the trees for the wood. Many people, from 1788 to the present, who looked at the Australian bush, expected to see a wilderness, and by wilderness they meant thick trees and thick undergrowth. There seems to have been little recognition that more open 'park-like' areas can occur as a natural consequence of combinations of soil and topography and climate, and no recognition that they can occur as part of a natural succession cycle following both fire and other environmental disasters. There has certainly been no recognition that the fire could be natural.

Just because the bush had more open areas, either permanently (as a result of topography etc.) or temporarily (as a stage in succession), means nothing in terms of Aboriginal actions. But this variety, this natural mosaic, would have been seen as the way things were, the way they were meant to be, and the way they should be kept. The idea that people would set out to remove all the rest of the trees to mimic the bare patches would have seemed as bizarre to Aborigines in 1788 as it does to conservationists today.

Australia is the only place in the world in which the people of a whole continent practised sustainable maintenance. They did so because, intellectually or instinctively, they recognised that, like Antarctica and the tropical rainforests, the environment of Australia was such that it needed tender loving care to survive. Aborigines were not conservationists in the modern Western sense. They had a job to do and a living to earn and kids to raise, on a continent in which on average these things are harder to do than anywhere except Antarctica. But they were conservationists in the sense of being conservers, people who recognised that, if you were to go on earning a living in this unforgiving land, you had to do your best to maintain it the way you found it. If you did that it would go on sustaining you, and for 50 000 years the maintenance of Aboriginal society has demonstrated the wisdom of this approach of reciprocal obligation between land and people.

Recent years have seen less wisdom, and the use of the Aboriginal environmental record, inconceivably, to support environmental change and development. How could this be so?

8

Convict's Dilemma

Environmental damage is caused by a form of the prisoner's dilemma, except that it is played by many players, not two. The problem in the prisoner's dilemma is to get two egoists to cooperate for the greater good, and to eschew the temptation to profit at the other's expense. Environmentalism is the same issue—how to prevent egoists producing pollution, waste and exhausted resources at the expense of more considerate citizens. For every time someone exerts restraint, he only plays into the hands of a less considerate fellow human being.

—MATT RIDLEY, 1997[1]

There has been much speculation about the reaction of the Aboriginal inhabitants of Sydney to the arrival of Governor Phillip and his unlikely crew of nation-builders. Candidates for Most Unusual Spectacle would include the ships, the guns, the white skins, the clothing, the flogging, the hanging, and the appalling food the colonists had to eat. All of these could have been rationalised away or equated with things and behaviours familiar to Eora and Dharug. The behaviour that the Aboriginal people would have found most inexplicable, and unforgivable, was the cutting down of trees and the clearing of acres of ground.

The most familiar sound in the first year (and beyond) of the British colony in Sydney Cove would have been the sound of axe and saw. Australia Day 1788 began the clearing of the Australian continent, and one of Phillip's first orders was that 'convicts would be landed each morning to dig sawpits and fell trees'. It is likely that these were the first trees that had ever been felled by humans in Sydney Cove, and, as the sawpits ran hot, the rapidly extending area of clear land was the first to be created by human activity. If Australia Day was moved from 26 January, as has been suggested, the day could be set aside as 'National Tree Awareness Day', or perhaps just 'Environment Day', to focus on the need to reverse the chain of events that Phillip set in motion in 1788.

The area of Australia covered by trees has been both considerably greater, and somewhat smaller, than it was 200 years ago, as the arid pulse of the continent beats. The genius of the Aboriginal economy was to extract the maximum amount of sustenance, for a large number of people, from a landscape with the maximum number of trees possible in a particular environment. The genius of the Aboriginal land management system was to maintain a large number of people, through times of changing society, culture and technology, while maintaining the plants and animals they shared the country with. For the last 10 000 years or so they have been a changing people in an unchanging environment, unchanging because they made it so.

If Australia was a giant clock, where wheels moved in order to move hands that would after twelve hours return to the same spot ready to start again, then the Aboriginal people were the watchmakers who kept the works clean and oiled and repaired to ensure that this kept happening in each area they had responsibility for. This is the genius of the Aboriginal social organisation: responsibility for the environment rested with those who lived in and relied on that environment. And that responsibility was clearly defined and, coming with birth, impossible to escape. Not only were areas of land defined for maintenance, but all elements of the environment—plants, animals, physical features—were also assigned. This may be one of the lessons we could learn from a group with years of experience in running the Australian system. The Landcare groups are a start.

But the chief lesson we need to learn is how to stop thinking of development as a major purpose in life, to recognise that growth cannot be maintained indefinitely and that sustainable development is an oxymoron.

The Yarralin people see the 'nurturance of all life as the ultimate achievement of the Dreaming' in which the desirable universe is in 'a steady state in which all life is maintained at optimum levels of productivity'.[2] An old Aboriginal man once said to W. E. H. Stanner, 'White man got no Dreaming'.[3] After 200 years it is time we had, or the new steady state of the Australian continent will be the steady state of Mars, in which the main interest is the colour of different rocks, and Sojourner's tracks in the dust.

One of the things we could learn from Aboriginal attitudes to the environment is to appreciate that a good human environment is one that contains all the living organisms it is capable of having. Aborigines would not have been able to conceive of a world without other creatures and plants. A rich life, a full life, is one shared with the rest of the natural world, not one in which everything is subsumed purely to human interests, and where other organisms that get in the way are removed, and where the environment can be damaged uncaringly and casually.

If you go far enough back in time, genetic relationships become increasingly close and strange. It is said, for example, that tens of thousands of British people are related to William the Conqueror.[4] Go further back, and all human beings are perhaps descended from a single woman in Africa. Back further, and the DNA we share with chimps and other great apes—over 90 per cent—reveals that we are all one family of cousins. An unimaginable number of great-grandparents ago and we share an ancestor with all those other animals with warm bodies, hairy coverings and mammary glands, many of which we eat. In fact, go far enough back and a single-celled organism crawling through mud 4 billion years ago is in the family scrapbook of not only every human family, not only every mammal, not only every other animal, but every plant as well. The Aboriginal world view, then,

144 THE PURE STATE OF NATURE

is accurate; humans are not different, not set on earth to have
dominion over other living things, but are just part of the universe
of other living organisms. People, animals and plants all merge
in the Dreamtime.

There is a conversational game in which you muse about
which figures from history you would go back and assassinate in
order to prevent bad things happening. People generally say banal
things like Hitler, but I have a very long list, mostly under a
religious heading. A variant of this game might be to see which
sentences written or spoken in history you might delete to
prevent bad consequences. I think there are two that leave the
rest for dead: 'And man was given dominion over the animals
etc.' and 'Go forth and multiply'. Paul Ehrlich, in one of those
sentences from history you'd love to have written, dealt with the
latter when he said 'no matter what your cause, unless you solve
the population problem, forget it'. Maybe an Aboriginal Ehrlich,
50 000 years ago, came to the same conclusion—but Aborigines
acted upon it.

The dominion business is as serious a problem in the damage
it has caused to the environment as is the population explosion
(like Bill Clinton's message about the economy, religious people
appear to have a similar message: 'what is the only thing that
matters in human life, what is the only aim? It's reproduction,
stupid'). The idea that the human race is the only important thing
on a planet with billions of years of history and millions of other
organisms is collective egocentrism gone mad. Instead of learning
to live with other organisms, thereby maintaining the diversity
that is so crucial to the survival of the world, there is a strong
strand of human thinking that says that anything that causes the
slightest problem, or even a mild inconvenience, must be
removed. When we are developing, nothing must get in the way.
This is a kind of ethnic cleansing on a worldwide scale—cleansing
the world of everything non-human.

It is a philosophy that will see the extinction of countless
species. It is a philosophy that becomes ingrained in children
because we continue to allow barbarities like duck-shooting and
shark and other big-game fishing. 'Shooting little birds' instils an
attitude that animals and Nature are just there at the whim of

humans, with no value of their own, and no right to existence if a human being decides to point a gun through a window and end their life.[5] It is an attitude seen in the horrors of Auschwitz, Bosnia, Rwanda and even Australia—the enemy, the 'other' is sub-human or non-human, and can be destroyed without thought. It is also the attitude that a death penalty instils in a society. America kills thousands of its own citizens each year, partly because of the insane gun ownership ethos, partly because of the belief in hunting as a test of manhood, but also because a death penalty sends such a clear message to citizens: it is okay to take an adult human life if someone decides that it is. Once you allow that state of mind, it isn't just learned judges who have power of life and death but every punk on the street with a cheap pistol or stolen automatic rifle. Attitudes to the environment will only change when humans agree that not only is gun ownership wrong because it causes human deaths, but because it causes the death of animals.

It becomes ingrained through circuses as well. Those that have wild animals help to inculcate in children the idea that any species is there simply for our entertainment and has no other value. Its survival is at our whim. A child who believes that cracking a whip and making a tiger sit on a stand is an appropriate relationship between tigers and humans is a child who will later happily kill possums, or bulldoze trees, or allow a factory to pollute a river. It is interesting that the first announcement by one of the successful far Right candidates at the recent ACT elections was not about health or education or aged care or the environment but that he would try to repeal the law banning circuses in the ACT. The Right have long understood that if you train a child early enough to think in the correct way, then the natural order of things can be preserved forever.

Elements of the hard Right economic-rationalist minimal-government movement have in recent times turned their attention to the environment, perhaps believing that having 'solved' all the world's social and economic problems they will now show that eco-rats can solve environmental problems too. The things about the environmental movement that appear to really irritate the hard Right, apart from the 'blocking of development', is that it involves people acting not selfishly but for a common good, and that government is necessarily involved. This is anathema to people who believe that no government is good government, that people act only through self-interest, and self-financial interest in particular, and that the world is there only to be exploited.

If indigenous people are 'true conservationists', then it implies that, deep down, all humans are true conservationists and we need to recover that ethos. If they are not, then modern humans need to invent a way of being conservationists. The hard Right would propose that you can only conserve things you can make a buck out of, and you only protect things you own, and that government should get out of everything except defence.

We have a classic example of the 'prisoner's dilemma' in Australian conservation. It is the Toolache wallaby. The last of the species seen alive seems to have been one held in captivity in Robe, South Australia, in 1927. Attempts had been made in 1923 and 1924 to transfer the last known band to Kangaroo Island but (in a phrase that still sends shivers down my spine) 'as a result of overmuch driving the few examples obtained were either dead or dead shortly after capture'. Then 'owing to the extensive publicity given to the two expeditions . . . realisation of the great rarity of the wallaby . . . survivors of the 1924 attempt have been wantonly killed for the sake of the pelt as a trophy'.[6] Not by Aborigines, of course—whitefellas with guns, operating in a freewheeling capitalist society with no environmental protection. There are other examples of the fallacy of the 'privatise it, put a value on it, and it will therefore be protected' theory—rhinoceros horn, elephant ivory, rare cockatoo species. As the exploitation nears its end, the value on the last remaining specimens is so high as to justify anything to get them. No matter whether someone owned it or not, the pressure to get it will override the ability to protect. Nor will the owner have any incentive to increase numbers. If the value of ivory is strongly linked to its rarity, then the private owner would be foolish to push the numbers up higher. There is a story about someone who owned the last two of a class of immensely valuable objects and deliberately smashed one in order to make the value of the remaining one not just twice as much, but infinite. The only course is to let no one have ownership but let everyone have ownership, and to let the species have zero legal value.

(It is all reminiscent of another of the great divides between liberal and conservative approaches to a pressing social issue—drugs. Conservatives believe in making war on drugs, cracking down harder and harder, stopping more shipments, jailing more dealers and users. But it is the same problem: the more successful the war, the more it contains the seeds of its own defeat, because the value of the drugs rises higher and higher with each 'victory'. The higher the value, the more impossible it will be to stop the trade. Legalising drugs removes the value and takes away all incentive for wicked men to profit from misery.)

The same relationship between scarcity, value and the drive to exploit applies in conservation matters. To hunter-gatherers, there is no 'rarity value'. The last specimen obtained is no more valuable than the first. Because there is no private but only group ownership, no one stands to make a profit. In fact, the reverse holds true: the last few specimens of a species will be so hard to catch that they have a high negative value, and will live to reproduce another day, while a more abundant species comes to be the focus of exploitation for a time. It is one reason why hunting 'overkill' as an extinction mechanism by hunter-gatherers makes no sense. Overkill by the entrepreneurs, the Buffalo Bills of the wild west (whether in America or Australia), does make sense, and is not a mechanism that is going to preserve anything more than the stuffed pelt of the last of each species as they become extinct like stars blinking out in the sky.

The idea that conservation is problematic because it lessens the 'value' of species in dollar terms, and that farming crocodiles for handbags is a model for the conservation for all species, is one that keeps popping up in strange places and wins great support from big business and from the Right of politics. It isn't hard to see why. It is an idea that fits with 'user pays', 'work for the dole', private education, HECS, private health cover, non-union labour, corporate arts sponsorship, advertising in non-commercial broadcasting, and so on. That is, nothing has an intrinsic value, the only reality is a dollar value; if you can't pay the fare then get off the train. It might also be suggested—cynically—that big business sees a chance to make some very big dollars indeed if no part of the environment is excluded from commercial exploitation. What is the background to the promotion of the idea of species having to pay their way or become extinct, and what would be the outcome of such a policy?

There has been some speculation about how the first Aboriginal people would have seen the Australian environment. Did they see it as an alien place that involved considerable change in their culture, and a steep learning curve to adjust to the new resources? The question really arises from a concept of present-day Indonesia and present-day Australia, and the obvious difference between, say, Sumatra and the Australian desert or the south-east. But even today there is no sudden contrast between the north of Australia and the southern part of South-east Asia. The genius of Alfred Wallace saw that Australian flora and fauna were not unique and totally isolated as they are generally portrayed, but Asian species gradually reduce in number as you approach Australia, and Australian species gradually drop out as you leave Australia. The area of transition is known as Wallacea, and this is the area where travellers from the north would first, and gradually, have encountered possums and wallabies and other marsupials, as well as Australian birds and reptiles and plants and so on.[7] At the time of the arrival, the wetter conditions of late Pleistocene times would have further increased the familiarity of the Australian continent.

Once people were on the Australian landmass there would have been the need to learn about new plant and animal resources, learn to avoid poisonous plants or how to prepare them to make them edible, learn about seasonal variations in abundance, learn about animal migratory and other behaviour patterns. But like their gradual and gentle geographic introduction to the country, such learning could take as long as it needed to. There was no rush, no hurry, no desperation; each generation could learn something new and pass it on to the next. With 3000 generations to play with you can pack a lot of expertise into people. By 1788 Aboriginal people had had a very long time to get used to

Australia—to them it was just the way a continent should be, and the plants and animals were just right. Sure there had been harder and softer times in the past, but there had been some 500 generations who had lived in the environment the way it was in 1788. It was a pretty extensive training program.

The British colonists who arrived in 1788 were part of a redeployment exercise, and their training had been very poor indeed. For thousands of years their ancestors had been progressively turning the British countryside into one that was almost completely managed and manicured. Wilderness was a sign of lack of care or of failure. Even the forests were extensively managed, and the rest of the country was farmland. Wildlife that got in the way of farming, like wolves and bears, was eliminated; land that was unproductive for farming was drained or cleared. The animals that were seen in the countryside were domestic animals; even deer were mostly in domesticated flocks roaming in man-made parks. The odd remaining wildlife, like badgers, foxes, hares, rabbits, was hunted extensively. This is countryside in which everything has a value, but it has little diversity, few features that are not the result of human intervention, and there has been considerable loss of species. It is countryside that is also familiar across much of Europe and around the Mediterranean. Even small birds have value in France and Spain—so much value that they are shot and eaten whenever possible (as they were in Britain well into the nineteenth century and even more recently—it is odd how quickly we forget our own cultural history when condemning others).

The colonists arrived in Australia to find a whole continent of wilderness, a continent of apparent failure to manage the countryside. It wasn't what they had been trained to see as a proper human environment, and they set about putting it in order, rather as one might start cleaning the rubbish out of a derelict house. And there was a lot of rubbish—kangaroos had to be replaced with sheep and cattle, valueless eucalypts had to be replaced with oaks and poplars and willows, the only large carnivore (the thylacine) hunted, like the wolf, to extinction, and so on. Every tree that remained alive was a few square yards less of grass or wheat. Rivers that ran to the sea were wasting their water. What appears to be a new idea about only keeping things that have human value is actually a very old one that goes back to the economic rationalists aboard the First Fleet. The problem with Australian conservation is not that the idea hasn't been tried, but that we are looking at the result of 210 years of the application of the idea!

Domesticating the environment means that at best only a few species of each kind of animal will be selected for saving. Competitor species will need to be removed and any environmental factors that detract from commercial performance will need to be changed. In the natural environment, the interests of all the components are balanced against each other. Factors that promote the interests of one species at one time will act against it and for another species at a different time. Commercial use of the environment is the opposite of this—imbalance must be promoted; the interests of chosen species must be made as high as possible and be kept constantly high.

Once you start harvesting a species, whether black cockatoos or kangaroos, or maintaining a captive breeding population of pandas or Sumatran tigers, you are on the inevitable path to domestication. Just as in the development of farming, subtle selection effects begin to bite—a favouring of the individuals that can be handled more readily, or thrive on zoo diets, or are not worried by human observation, or are less susceptible to disease, or who produce more young in captivity, or have more attractive plumage or whatever. We don't think that having domestic sheep is a way of preserving wild sheep, nor should we believe in this prescription for other species.

Harvesting species also changes attitudes towards the wildlife concerned. This is no longer something running free; it is something we have control over, and therefore it is diminished in value. It is the same process as the use of animals in circuses, and the hunting seasons, whether on ducks or grizzly bears or lions.

Not only are competitors to be removed, but the imperatives of commercial development mean that there has to be a continual striving for more and more efficiency. If, say, kangaroos are being farmed as a way of preserving them, it would quickly become evident that one or two species were more profitable than others—faster growing, better behaved, better meat producers, faster reproducers etc. Those farmers with the less profitable species would have to shift from them and get into the other species. The same pattern is seen in our major domestic animals. In the sheep industry over the last few hundred years there have been major shifts in commercial needs and fashions and productivity. In England in the eighteenth century new breeding practices led to the development of sheep breeds with better wool and meat production. These breeds quickly took over and many regional breeds either became extinct or survived in small numbers as curiosities. Commercial imperatives don't encourage diversity of breeds.

They also don't encourage diversity within breeds. The selection and cross-breeding process in many breeds have resulted in loss of the original form and genetics of those breeds. In some cases there have been searches to try to find if there are any individuals of the original form of breeds still in existence. Other breeds are desperately trying to retain their purity. The final stage in this kind of development is the use of embryo transfer techniques, genetic engineering and cloning. Already there are breeds of cattle whose genetic diversity is being reduced by the use of individuals with desirable characteristics as the source of large proportions of the herd. This process is likely to reach an end point where every member of a herd is an identical clone deriving from a single individual.

The process is also well under way in plants used in agriculture and horticulture. Genetic variety is a curse of efficiency. A factory is efficient because of standardisation of parts and processes. We are headed towards factory farming quickly. But we have seen what it can do already. Ireland in the nineteenth century was a potato factory. The potato was easy to grow and it filled bellies, and it rapidly took over as *the* agricultural product of the country. But there was no variety within the potato in Ireland; it was a monoculture, and when the blight hit it wiped out the potatoes and almost wiped out the people. This is the scenario the new ecological rationalist potato-eaters have for Australian plants and wildlife.

In recent times conservatives have put a blight on the idea

that anything should be owned by the public, and the power failures in New Zealand and the gas disaster in Victoria are perhaps among the early signs of the effects of this policy. To the conservative mind—and again, I have trouble coming to grips with this—the ethos seems to be that because private enterprise aims to make profit, this will drive the economy up to the benefit of all. But they have lost sight of the profit equation, profit equals income minus costs. Sure you can raise profit by raising income, and this may (or may not) have a positive effect on the economy. But you can also do so, more easily in the short term, by lowering costs, and we have seen a savage wave of sackings and redundancies to achieve this, and deregulation to ensure that such trifles as safety regulations no longer need to be complied with.

This same approach would apply to making a profit from wildlife. In the Aboriginal economy, costs rise as things become rarer, and the response is to switch exploitation to another species or area. This works to prevent environmental damage or even extinction. In the Western economy, changing either part of the equation by increasing exploitation to increase income, or making changes to cut costs, would work against conservation.

It also works against the interests of the community generally, as we see in the headlong rush to increasingly privatise everything from power and water utilities to schools and hospitals, transport and communications, roads, sewerage, employment agencies (we can even see a dollar to be made out of the unemployed, as the owners of the nineteenth-century workhouses did). Public ownership developed to make sure that the interests of the public were protected. The rush to sell off everything will take us back to the early nineteenth century where the only motive is profit, and every aspect of life is there to be exploited.

In the eighteenth century, Aborigines had difficulty with the idea of private ownership they found among the colonists. In April 1788 Phillip 'signified to the convicts his resolution that the condemnation of anyone for robbing the huts or stores should be immediately followed by their execution'. There were numerous examples of Aborigines admiring various articles and taking off with them, which outraged the colonists. The Aborigines would have seen the objects as communally owned (particularly, perhaps, since the colonists were occupying their country). It took a hundred years before communal ownership in the non-Aboriginal society of Australia began to develop and catch up to the Aboriginal way of doing things that had served the country so well for so long. It has taken frighteningly little time to turn the clock back.

The past, whether thirty years ago in Adelaide or 30 000 years ago at Lake Mungo, is shadowy and elusive, streaming away like tendrils of mist as we try to grasp its reality and its meaning. People everywhere, at all times, have mined the past for their own social, political and cultural purposes. Many people mine the past to get clues about the likely course of future events; others, much more dangerously, do so to try to change the future. But preaching to the current generation about the future is a very uncertain proposition when your pulpit sits in the past. What appears to be a high, rocky mountain top from which you can spread the word could quickly turn into shifting quicksand.

People of all religious persuasions have a desperate need to treat scientific facts (about evolution, cosmology, geology, physics, behaviour) as if they were theories, desperately trying to give the illusion of uncertainty where there is certainty, in order to fill the past with mythology and supply yesterday's answers to today's questions. Politicians have the opposite problem, too often treating scientific theories as if they were facts, desperately trying to create certainty where there is uncertainty, to give the illusion that today's answers can be supplied from yesterday's questions. In fact, we need to understand the questions of the past—they remain good questions today. For 50 000 years Aborigines had to deal with one big question: what is the best way to conserve the Australian environment? It remains the big question today.

The question of what caused the extinction of the Australian megafauna is one unlikely to be settled archaeologically. The question of the effect of Aboriginal use of fire on the Australian environment is perhaps even less susceptible to solution by archaeological techniques. The data available are about as good as those relating to the effects of immigration on employment, or the chances of success of different approaches to drug abuse, or the role of Greenhouse gases in climate change. What people believe about such questions is a measure of their political stance. What they believe about extinction and fire is also a matter of politics. But the questions are worth asking, if only to underline the nature of our choice for the future.

Australian Aborigines are crucial to the debate about conservation and the environment. Of all the indigenous peoples of the world they are seen as those most closely attuned to Nature, with every aspect of their society and culture and religion intimately bound into the rest of the natural world. In the past Aborigines were used as an example of how we should behave towards the environment, an obstacle to those who wanted a licence to destroy the environment for gain while claiming that this was natural behaviour. Now even this tiny obstacle to development has been removed by a mistaken view of the past proposed by those who think Aborigines caused environmental damage. If even Aborigines caused massive environmental change, then the human race has no natural instinct for conservation. If even the Australian environment has been moulded by human activity, then there is no need to conserve anything except what we choose to conserve because it has a monetary value to humans. On the SBS 'Insight' program in June 1999, Michael Archer commented that we should commercially exploit Australia's fauna because it would make us 'filthy rich'.

One of the most crucial questions ever posed about Australian conservation is that posed by Rhys Jones when he asked whether, had humans not discovered the Australian continent until 1770, there would have been Diprotodons wandering down to the water's edge of Botany Bay to be described, matter of factly, by James Cook.

If you believe that, had it been truly *terra nullius*, Australia would have been populated by megafauna instead of people and would have had a considerably different pattern of vegetation, you will be inclined to believe that this gives the human race a licence to keep indulging in massive environmental change. If you believe that the megafauna became extinct as a result of climate change, in spite of every effort by Aborigines to maintain the environment unchanged, and that the vegetation was in a complex equilibrium with soil and climate and topography and in turn with a natural fire regime, then you will be inclined to believe that we as a people must do our best to conserve what we have for future generations.

The important lessons from Aboriginal people are that our home is not just a built structure but the whole of the environment, and that group action to maintain that common home, not self-interest, is the key to having a future. If we don't begin to truly understand the past, instead of inventing a past that suits a particular political and economic agenda, there is unlikely to be a future worth living.

Australia has two other lessons for the world. A major problem with environmental protection is the boiled frog syndrome. If a frog touches boiling water it will instinctively leap away. But if you put a frog into cold water and gradually raise the temperature, the frog will eventually die in boiling water without ever being aware that conditions had passed a critical point. In our environment, change can happen so gradually that we are unaware of it until it is too late (climate change resulting from Greenhouse gas increases is a classic example, probably literally, of the boiled frog syndrome). A few invading weeds, the loss of a species, a dying tree, a small amount of erosion, are all tiny changes that can result imperceptibly, over a long time, in massive changes. Old people remembering the conditions of their childhood, or people using old photographs (for example of the Barrier Reef) are cases where we suddenly become aware of the effect of accumulating small changes.

The Aboriginal approach to this problem was one of extreme conservatism. They had instinctively realised that environmental change was a slippery slope. Their technique was to try to prevent any long-term change. While there would be seasonal variations, and bad years, the aim was over a period to begin a new cycle at precisely the same point and start all over again. We too need to be alert for small changes, and aim to maintain the environment at its present point (or better).

The consequence of this process by Aborigines was that effectively they created a situation where the continent would have been the same whether or not they had been here. The idea of maintaining the environment is rather like the attitude of responsible campers in a national park—don't damage things, don't leave things behind, leave the site exactly as it was when you arrived. A lot of work goes into maintaining a campsite in its original state. But *this history* has been against the interests of Aborigines in recent Australia. They have been so unobtrusive in their efforts that they were seen as mere parasites, rather like the situation of a hard-working but quiet schoolchild.

Conservatives have found a new catchcry: 'zero tolerance policing'. It is inappropriate and extremely damaging in law enforcement, but what we need is Zero Tolerance Environmental Management as practised by Aborigines. It is so easy to say that some small change doesn't matter—the removal of a tree, the damming of a creek, the introduction of just a few more sheep into a paddock. But each small change brings us one step closer to a big change. A Zero Tolerance approach would make us alert for those seeds of destruction—the butterfly wings flapping in China causing dust storms in Melbourne.

The second lesson is the life raft idea. To Aboriginal Australians, Australia *was* the world. There was nowhere else to go, and therefore you had to maintain the environment, and control your population within this continent, or you were doomed. Other civilisations have seen the whole world as their playground. If things got tough, or too crowded, you simply moved on somewhere else. There was always new ground (with or without indigenous people) to exploit, and the idea of an ultimately limited carrying capacity for the planet as a whole was (and is) impossible to get across to most people.

The Gidjangali of Arnhem Land and the Alyawarra of the desert provide a classic example of the Australian dilemma. You can have few people relative to your resources and lead a good life with choice, or you can have many people relative to few resources and lead a life of scratching for a living, with no choices. There are still people using the nonsense term 'sustainable development' (by definition, no development is sustainable), or suggesting populations of 12 billion for the planet, and of 100 million for Australia, as being not only feasible but desirable. The lack of imagination about what such populations would do to this continent, and to the planet as a whole, is staggering. Even more staggering, such proposals come from people who don't have the imagination to see the potential effects of Greenhouse gases, or forest clearance, or pollution over a long period.

Another aspect of the Aboriginal genius is that their religious system was compatible with the idea of maintaining population at a low and sustainable level. In a sense, large numbers of people conferred no benefit; it was the maintenance of *categories* of people that was the important thing to keep the system going. The fact that your role in maintaining part of the land was determined by attributes of your birth, and that conversely every piece of land had a corresponding category of person, meant that even with very few people the whole continent was cared for on a continuing basis, and culture transmitted from one generation to the next. Many other religions have taken the view that the more people, and therefore the more souls that finish up in the particular heaven concerned, the better. It is a view that will ultimately cripple the planet, with many people living in the resulting hell on earth. It is related to the view that life on earth is of little, if any, importance or concern, and that some afterlife is the only reality (a view taken to its most extreme form, perhaps, by the poor crew of California's dreaming, shucking off their bodies as so much rubbish on spaceship Earth, while their souls winged their way to join up with another spaceship, trailing Hale-Bopp, to take them away from all this). Aborigines are concerned with the here and now, making life as good as they could before reabsorption into the land. It is a pity isolation stopped Aboriginal religious belief being exported to become a major world religion.

We have to treat the whole planet, isolated in a cold, hard universe, the way Aborigines treated Australia. You can't be taken away from all this—there is nowhere else to go.

9

Ghosts

The most controlled understanding of [time] is by reckoning in
terms of generation classes, which are arranged into named and
recurring cycles. As far as the blackfellow thinks about time at
all, his interest lies in the cycles rather than in the continuum,
and each cycle is in essence a principle for dealing with social
inter-relatedness.

—W. E. H. STANNER, 1967[1]

There is a game that I played as a child called Ghosts. In Ghosts
someone is blindfolded in the middle of a circle of children who
try to creep forward without the child in the middle hearing
them. If you hear something, you point to where the sound came
from, then that person is either out or must go back to the
starting circle drawn in the sand. Eventually someone manages
to creep up so silently that they can reach out and touch the
blindfolded child, and then they swap places. We are all blindfolded, listening for those tiny movements. If we don't hear
them—and they are imperceptible—we will be overwhelmed in
a final rush. This is much like the reality of environmental
damage. Some players hear very well, but others appear unable

to use their sense of hearing, and are deaf to both the reality of environmental change and its consequences.

I can no longer play Ghosts; an old man's hearing gradually fails. But my environmental ear is working better than ever. Perhaps Environmental Ghosts is a game for the elderly and the elders. A game where you have sufficient span of memory to remember the world of your childhood and how it has changed. A game where the old men of the tribe remember the sparrow falling and the tree burning, and can warn the young that adjustments need to be made to take the country back to the status quo and begin the march from the edge of the circle all over again.

When I was a child my friends and I marched up to the top of the hill to school each morning and back down again in the afternoon. Imperceptibly our feet moved the grains of sand and limestone, shifting them, rolling them, and then letting them blow away or wash away. The tiny groove that our feet began those forty years ago is now a metre-deep trench, and children who could be my grandchildren trudge through it, having no concept of a time when it was not there.

My grandmother had an egg-timer, designed like the old hourglass, but so small as to run for only three minutes while an egg boiled. It was a vestigial piece of technology, last remnant of a mechanism that was once the chief means of telling the time. I marvelled at it, wondering how someone had known exactly how much sand made three minutes, wondering too how it came to be so exact. Clockwork was one thing, but sand running through a bottleneck? You turned it over and the miracle began again, over and over: empty became full, full became empty, as often as you cared to set the clock back to zero.

Buckets of ashes tipped from a fireplace every morning for ten years can create a large midden. Three pairs of leather-soled shoes scuffing the surface of the ground for forty years can dig a deep hole. Even imperceptible effects can start to assume great magnitude if repeated often enough through the long period of the average life, the great period since 1788, and the immense period since the first occupation of the Australian continent.

I sit looking out of my window again as I write. The misty green of winter has given way to the browns and yellows of late spring and early summer. The quietness has given way to the roar of the wind, in a typical blustery spring day on the southern tablelands. Where we removed the grass to build the house the dust is moving, and then the topsoil, in great streaming clouds. As I sit on a bare patch of ground, I watch the dust blow towards the Tasman Sea, individual particles in a great dance of time that will eventually lead them away from me to the bottom of the sea a thousand miles away. Down in what was a small creek, home to late-hatching tadpoles, a last drying, gradually shrinking puddle will be their grave as the last water boils away from the hole before they have time to become frogs.

I remember that terrifying image a few years ago of the great dust cloud, like mustard gas in 1917, rolling towards and then engulfing Melbourne. That cloud too began with small bare patches. All the hottest years on record are in the 1990s, they say. El Niño strikes the country again in a fury exacerbated by the Greenhouse effect, but our politicians argued in Kyoto for more, much more, of the same. The sand grains are moving and the frogs are boiling in their waterholes.

Ghosts people our individual pasts and our collective past. We can try to bring those ghosts to life, but we will fail; attempts to conquer history, to bring certainty to the transactions of the past, are doomed to fail. The past is another country, a country of leasehold title, which we are free to use but not to own. No one owns the past, no one has exclusive use of it, and no one can be certain about it. But we can do our best to interpret the past wisely and use the lessons well.

And we can try to conquer history as the Aborigines did and many still do, to be aware of subtle changes and their implications and to aim for constancy, for the retention of a lifestyle they enjoyed in an environment they loved—not an unworthy aim.

We must embrace the past, feel the structure of it in our bones, as we feel the bones of the land through its contours. The past has its own contours and rhythms, and until we can immerse ourselves in it, truly feel it, see the ghosts through the trees, we will not understand it. It is not a question of conquering the past but accepting it. Those who believe that the past can be con-quered, that history can be ignored as bunk, that its lessons have no meaning, that the past is always with us to be constantly revisited and reused, are dangerous for our future.

We have recently, as I write, celebrated another Australia Day, a day in which the ghosts of redcoats wandered the Sydney streets, the sound of fireworks echoing the sound, perhaps, of earlier volleys of guns fired in celebration and guns fired in anger at Aborigines disappearing like ghosts through the trees. The redcoats conquered history, ending 50 000 years of one kind of history and beginning another. Some of their descendants conquered that history by denying it, changing it, rewriting it. To others, the Aborigines were always invisible in *terra nullius* and should remain so—witness the outrage from the far Right when Cathy Freeman was named Australian of the Year.

But we can also conquer the past if we mythologise it, inventing golden ages that never were. Myths about the past may give comfort to individuals, people, and countries—we all get the histories we invent. But all myths are likely to be replaced by other myths, those of less comfort. There is little defence against this except the honest endeavour to conquer the past by recovering as much of the truth from it as we can.

The Australian government's position at the Kyoto Greenhouse conference in 1998 (and the lead-up to it) outraged conservationists. I felt the outrage also, but there is little point in outrage about this specifically. The position is no different from anything said by the conservative side of Australian politics for many years, and indeed is no different from the position of the conservative side of politics in most countries around the world, including the USA. It is a position people with this world view (a term impossible to use except ironically) will continue to hold into the future, while the dust blows, the forests disappear, the temperature rises, the storms increase in ferocity, the droughts gain in intensity, the seas rise, biodiversity becomes only a theoretical concept, people die by the millions, and the survivors live a life of an unimaginably poorer quality than our own. Because this result seems so bleeding obvious, the attitude of the conservatives is absolutely baffling. It is worth examining how it is they can hold these views, and then attempting to counter them.

It is not just a world view of conservative politicians pandering to their constituency of big smokestack business and 'we need to clear more trees to catch up to the other states' graziers. It is also a view held by conservative commentators. A recent one I read took great glee in pointing out that some particular bird species said to be under threat of extinction had survived, proof that greenies were scaremongering and, by implication, nothing was really under threat, and we should just get on with development and ignore them. It is an argument like that of fundamentalist religious believers who, reading a debate relating to some minor detail of evolutionary theory, choose to believe that any debate is a sign that the theory is wrong. Why the blindness?

Is it a combination, I wonder, of a lack of imagination, an inability to visualise alternative universes, and a belief that the only good landscape is a managed (controlled) landscape? My great-grandfather, Robert Charles Young, would have subscribed to that view as he manicured lawns and raked gravel drives, and planted herbaceous borders. In both respects it is instructive to look at the corresponding Aboriginal world view. It is a commonplace among anthropologists that Aborigines believe landscape must be managed —that, in effect, the concept of wilderness would be as much anathema to Aboriginal people as to the most extreme right-wing member of the Cattleman's Union.

This is the fundamental core of the firestick farming hypothesis. It is not just a colourful image, but is intended to be an exact descriptive term. Certainly there is endless ethnographic material relating to the use of fire to 'clean up' country and the like. But this is a misunderstanding of the Australian landscape itself, a misunderstanding of Aboriginal views, and a misunderstanding of the nature of farming. Farmers aim to maximise the return from a piece of land. All activities should be geared towards extracting the greatest possible income from the lowest possible investment, though in recent years there has been recognition by some farmers that there needs to be some easing back from pure economic rationalism in order to keep the system functioning. The environment seems curiously unaware that economic rationalism is the only valid approach to life.

This is the reason why so many trees were cleared from farm land. Every tree was seen as reducing the amount of pasture, and therefore the number of livestock, or the amount of wheat. Kangaroos and parrots are shot because they eat grass or wheat, eagles and crows for their attacks on sheep. Native grasses were removed and replaced with a small number of introduced grasses, heavily dependent on fertiliser. People no longer live off the produce of their farms, which are specialised factories for meat or wool or wheat. Any environmental diversity is a threat to the efficiency of those factories.

Aboriginal people also wanted the maximum return from a piece of land, but the maximum return came from having maximum diversity. Because a group had to find almost all its requirements in its own country, it was utterly important to maintain the diversity that gave you red meat and cereal and birds and fruit and roots and shellfish, as well as spears and housing and clothing. If the Aborigines were farming, it was farming more like the practice of hundreds, perhaps thousands of years ago than anything in modern times—though even last century, parts of Britain and Europe had small holdings where people endeavoured to produce most of their own food. My great-great-grandfather, on just a few acres, grew wheat, fruit trees and vegetables, had cows and chickens, and provided most of the food for the family from what he produced.

The conservative mindset seems to lack the imaginative ability to picture alternative worlds, alternative futures, alternative views of the past. They appear to believe, like the worthy gentlemen of Victorian England, that everything is for the best in the best of all possible worlds. It certainly was if you were a worthy Victorian gentleman, but not if you were a peasant, one of the urban poor, or black. The best of all possible worlds can only exist if it is so for every human and for the environment. In the progress-obsessed mentality of our forebears, nothing could be allowed to get in the way of business, the environment was there to be conquered, unionism was forbidden, education and health were for the rich, the workhouse and early death awaited the poor. Until recent times, however, we have shared the belief that things were gradually improving—not reaching a peak in late Victorian England and then stopping, but continuing on for another hundred years. It is an idea of evolution carried into social and cultural matters. You made advances, then you built on those advances for the next improvements. It is a view of the world as an ever-expanding universe.

An example of an advance was the attitude to the environment. In the 1960s and 1970s there appeared to be a growing awareness that the environment was fragile and was being damaged, and that both population growth and development had to be brought under control or we were headed for disaster. But the present inability to picture things being any different—as a result of Greenhouse or population growth or pollution or land clearance—and the belief that this time, now, is the best of all possible times and the world is to stay as it is forever (conservatives, of course, see themselves as being unchanging people in an unchanging land), have sent us back to the start again. Back to the world of Victorian England and America, where pollution poured from factories unchecked. Back to the past in Australia, where the tree stood in the way of progress and had to be cut down.

For a while, just a while, the idea that Aborigines had managed the land, and that it was very fragile, started to have an impact. But in recent years that approach has backfired. Extraordinarily, Aboriginal land maintenance has been used to justify continued large-scale land clearance and control burning of the landscape. Extraordinarily too, we have gone back to a notion of *terra nullius*; the ghosts of James Cook and Arthur Phillip rise again to claim the land for the monarchy. So it is a lose-lose situation: Aborigines are supposed to have worked the land in order to justify the continued burning and clearing of the bush, but not to have worked the land when it comes to native title.

In health, in education, in workplace relations, in indigenous affairs, in the management of the environment, we have to start all over again, like Sisyphus, starting from the bottom of the hill to roll the stone up again. The view, when you reach the top, is much better. But the danger for the environment is that you can't roll the stone from the same place each time. Each time you lose a bit, more grains of sand rolling down a hill. If you carelessly lose plant and animal species, you can't resurrect them; lost is lost, gone is gone forever. If you want the land to be truly unchanging, then you must mark the fall of the pipit, and sense the movement of grains of sand.

There is another reason, it seems to me, why the conservatives are so opposed to environmentalism, so certain are they that the future is fine as long as we return to the values and behaviour of the past (or at least their image of the past). There is a clue in the attitude of Prime Minister Howard towards the Stolen Generations, and his refusal to apologise. Why not, you may ask. An apology costs nothing. But it does, you see. It costs part of the vision that these people have about the Australian character, the Australian ethos, the Australian history.

In the mindset of the Right of politics in Australia, the past is a golden age (always to be compared with a fearful present and a potentially frightening future of more change), a time of heroes and saints, selfless acts of comradeship and sacrifice. Our soldiers are always heroes and never commit a bad act in war; the explorers, the settlers, even the bushrangers (though this is a special case), the missionaries, former (conservative) prime ministers, are all good and great. Our sportsmen execute great feats of skill and endurance, and never cheat—they 'nothing common did or mean'. Our farmers developed a wasted land, created wealth and abundant food out of wilderness, and certainly caused no damage. At the heart of John Howard's sense of the nation's identity, and through it his own fundamental core of identity, is this belief that there is only good in Australia's past. Strike at this belief, cause a crack in the foundations of Australia's image of innate decency, and you strike a blow at the self-worth of the Prime Minister and all the millions of other John Howards across Australia. No matter what the evidence, he cannot say sorry and admit wrongdoing. But it is time we recognised the truth, apologised to the environment and then set about trying to put things right.

The hard-faced men and women in government, in the boardrooms, in the upper echelons of the public service deal in certainty and control. The idea that the world is beyond their control is at odds with their sense of identity and self-worth. Furthermore, the conservatives believe, or profess to believe, in individuals. The word 'public' (as in public service, public health, public education) has become a term of abuse. And there is nothing more 'public' than the environment. Unless people can act together, unless there is regulation in the public interest, unless there is control of rampant capitalism, unless there is a diminution of individual self-interest in the cause of the public good, then the world is doomed. Fortunately the pendulum seems to be swinging back a little from the Right. The recent elections in Germany, and the resulting incorporation of Greens in government, and the repudiation of the hard Right in American mid-term elections in 1998, are both cause for some guarded optimism.

But Australia has only recently swung to the Right, and a lot of damage can be done yet. As Big Business has regained the ascendancy it only ever loses temporarily in Australia, we again hear the increasingly strident demands that nothing must get in the way of business. In the last week, driving from country to city, I again hear that nothing in the local election is more important than the economy, and that any hint of regulation of development must be stamped on because nothing is more important than developers.

In the nineteenth century, unchecked private enterprise turned countryside into wasteland and cities into slums. Where profit for a few is the only motive there will be no profit for society or the environment. Gradual realisation of the damage that had been done, and recognition that the profit motive acting alone must inevitably cause damage, led to a slow development of regulation and law designed to ensure that government also worked for the rest of us. But the neo-conservatives have stripped away much of these balancing mechanisms, and are sending us helter-skelter back to the world of the nineteenth century.

In environmental issues as elsewhere, the cry is to get rid of the rules and allow unchecked development to suit the individual. In Aboriginal Australia we had the supreme example of a society where the rule of law was everything. From 1788 it came up against one where the interests of wealth ran rampant, and the selective application of the law to the poor and unfortunate could be seen as the ships unloaded their human cargoes.

In Aboriginal society the environment could be maintained for 50 000 years because there was an ethos of keeping it unchanged, a monitoring of change, a division of the whole country and assignment of group and individual responsibility for every part, and a strict and complex system of rules and laws and penalties making it all happen. There was a sense of custodianship, mutual obligation, towards the land that supported them. We could usefully emulate this.

The environmental movement is always seen as being for the young. They, after all, are the ones who will inherit whatever mess is being created now. But the young know little of the past before they were born, little of the way things were. We need elders attuned to the finest movements, the finest change, the smallest sound, who can warn and reset the clock. We need elders from all the continents, meeting together to stop the gathering momentum of change that threatens to engulf us all. We need the old people with enough span of years to remember the way things were and advise on restoration and repair. The elderly have been heavily involved in Neighbourhood Watch, that system of observers who know their community and know when something is wrong. We need a Neighbourhood Watch for the Land, a Green Watch, a system of local groups encompassing every part of the continent. Native title has been much discussed, but there needs to be a sense of communal title, in which damage to the environment damages all of us. The environment is the property of all of us.

Lloyd George spoke in 1919 of the conservatives on the opposition benches as being full of hard-faced men who looked as if they had done well out of the war. The current rulers of many Western nations, including Australia, have the look of hard-faced men who have done well out of the environment. The involvement of old people would have a further advantage. While the environment is a youth issue it can be ignored by the hard-faced men, dismissed as romantic nonsense because it is the idea of young people who know nothing of the economic reality of life and politics. As a movement of elders it would gain strength and prestige—witness the example of Dame Rachel Cleland, widow of a Menzies-era Liberal pollie, who at the age of 93 is out there with the demonstrators trying to save the forests.[2] Green Watch and Grey Power may become the combination from the past that saves our future. The old also have that sense of other possibilities, other worlds, other alternatives so lacking in the imagination of the hard-faced men. Aboriginal elders, telling the next generation of the other world of the Dreamtime, prepared them with an imagination that could encompass the idea of change in the environment and resist it.

It is said that those who do not understand the past are doomed to repeat it. Perhaps those like the Aborigines, who repeated the past, were destined to understand it. There is much serious work to be done to save the Australian environment, and the environment of the whole world.

I come back to firestick farming and the megafauna: the serious work that is required is totally obstructed by a reference to a past Aboriginal–environment interaction that never was. Australian prehistory and its interpretation have in recent years been hijacked by the Right. They need to be reclaimed by the Left. If we need a model for caring for the Australian environment, then it is not a time for outsourcing to Europe or America. We have a home-grown job applicant with 50 000 years of experience.

10

Theses nailed to the door

. . . for the purpose of eliciting truth.

—MARTIN LUTHER, 1517

When Norman Tindale decided that Aborigines had extensively changed the Australian landscape by the use of fire, so extensively that there was no natural vegetation left, he caused a great shift in the belief systems of Australian archaeologists and those elsewhere. It was, at the time, a welcome corrective to the view of Aborigines as 'parasitic', a view that began with James Cook and had continued into the 1950s and even today. *Terra nullius* meant, in effect, not making use of the land, and here was a massive antidote to that description of Australia.

The general public makes a distinction between religion, based on faith and belief, and science, based on facts and experiments. While it is true that there are no facts and experiments in religion, it is emphatically not the case that there is no faith and belief in science.

A radical new theory like Tindale's has the same impact in the scientific community that a new religion does in the religious community. Converts flock in, and what was once reformation becomes new orthodoxy. The converts, like Tim Flannery, see the whole world through the eyes of belief. Everything either becomes consistent or can be explained away. And eventually the new church closes in on itself, the doors are closed, and non-believers are treated with suspicion at best, at worst hate (and, in the world of religion, though not yet science, death to non-believers if the opportunity arises). It takes a challenge, nailed to the church door, to get a new view of the world going.

The strong element of belief is as dangerous in science as in religion because it prevents people seeing things. If you believe that people caused extinctions, and that if they did the theory requires that it happened very fast and left little evidence, then the less evidence, the stronger the belief. If you believe that Aboriginal use of fire modified the environment and turned forest into parkland and woodland into grassland, then every explorer's mention of trees more widely spaced than seemed natural to him becomes evidence for firestick farming. If you believe in firestick farming, then any change in the pollen record becomes evidence for human arrival, at no matter what date.

So let us try to nail some theses to the door, though I think we can do it in less than Luther's ninety-five:

1. You can learn from the past, but anyone who professes certainty about the events of the past, their cause and meaning, including me, should be treated with as much scepticism as someone who professes to foretell the future.

2. Aborigines haven't been here forever. But at more than 50 000 years, they have lived on the same piece of land longer than any other human group on earth except the Chinese.

3. Isolation, whether in Australia as a whole or in Tasmania, doesn't result in brain death. It does result in unique cultural and social attributes.

4. It is not because they were isolated and stupid that Aborigines didn't practise agriculture, but because the links between religion, society, culture and the land were so strong that they could not be broken.

5. Nothing—plant, animal, or human—is doomed to extinction by its genetic, taxonomic, social or cultural nature.

6. The species of plants and animals, their abundance, their arrangement into ecological communities, and the form and distribution of those communities, as seen by William Dampier, Jan Carstenz, James Cook, Willem Vlamingh, Tom Cobley and all, would have been the same whether or not Australia was or was not also occupied by people.

7. Aborigines did not cause the extinction of the megafauna (or the dinosaurs) and it is unlikely that they have caused the extinction of any element of the fauna and flora. The environment that present-day Australians inherited when James Cook planted the flagpole was the end result of millions of years of evolution and adaptation to a unique climate and land.

8. The climate we have today in Australia is a part of a cycle that is perhaps a million years old. Aborigines didn't change the climate, nor could they, but we certainly have the power to do so. This is the first period in history when humans have begun changing the climate of the whole world, and it will not be a change for the better.

9. Aboriginal use of fire changed nothing in the environment except in the sense of the short-term outcomes that follow any fire.

10. Aboriginal interaction with the environment was aimed not at exploiting it by changing it to a simpler form, but at maintaining the biodiversity at the highest possible level.

11. If you want to see what the Australian environment looked like in 1787, take a look at a wilderness area. It won't be exactly the same, because the effects of 20 million people and their baggage are all-pervasive, but it will be pretty close. All other parts of Australia are now very much depleted.

12. If you want to practise control burning in order to protect houses or farms, then do it in the same way as you would

use a bulldozer to clear a firebreak, but don't pretend that you are doing anything but damaging the environment.

13. Australian animals have become extinct in the last 210 years not because Aborigines no longer practise 'firestick farming', but because of, in order of importance, land clearance, feral animals (through both competition and predatoriness), and human predatoriness.

14. If you commercialise an environmental resource, you do so to make money. Don't pretend that it also benefits the environment.

15. The economics of hunting and gathering help to conserve species (and prevent extinctions). The profit motive does not (and can cause extinctions).

16. Aboriginal people weren't conservationists in the Western sense, but the effects of their beliefs were so strong as to protect the environment for a very long time. If this is true of Aboriginal people, effectively isolated from the rest of the world for years, it is a fundamental part of being human. Pursuing agriculture means you must suppress this human trait.

17. The Australian environment in 1787 was not an artificial construct of human making that needed to be constantly interfered with. It was a natural construct of a long history, and the things we are losing now will not be recovered.

18. You can't conduct experiments in the past or the future. Certainty of belief is dangerous because you can't readily undo mistakes, and much will be irreversible.

19. Such losses and changes in fauna and flora as Australia experienced in the past were the result of massive swings in climate. This is going to happen again with Greenhouse, particularly if this exacerbates the El Niño events. If we don't succeed in stopping this change, then we will again lose extensive elements of the environment. We will no more be able to prevent this effect than Aborigines were 25 000 years ago.

20. The more diversity we lose, the less chance there will be for environmental recovery ever to occur.

21. Maintaining trees was a vital part of Aboriginal maintenance of the environment. Australia now has only a tiny percentage of the number of trees present in 1788, a number that had been kept at the highest possible level through years of climatic change. It is not just the loss of forest areas that has been critical, but the loss of trees in the woodlands of the slopes and plains, once an enormously rich and diverse area biologically.

22. Societies that not only allow but encourage their citizens to slaughter each other are suicidal. Those that allow people to slaughter animals are equally suicidal.

23. There is nothing that is not related to the environment. We don't have a choice between, say, jobs and the environment, or education and the environment. Whatever your cause, forget it unless you look after the environment.

24. Until a future US presidential candidate (or an Australian prime minister) puts a message on the wall to remind himself (or, hopefully, herself) of what the only issue of importance is, and this message reads not 'The Economy, stupid!' but 'The Environment, stupid!', things are not going to improve.

The extinction of the Australian megafauna, the result of a climatic event, was a very loud early wake-up call. Now that it is high noon, have we heard the call? That change to a hotter, drier climate caused a large number of extinctions, even though there were no exacerbating effects of human modification of the environment, and it will happen again. Aboriginal maintenance of the environment, the retention of trees and biodiversity, is a greenprint for the future. If the effects of Greenhouse arrive when the environment is not equipped, then the effects will be greatly multiplied. We need to be in shape to weather the storm, just as, we are told, the economy needs to be in shape to weather the Asian economic crisis. The coming environmental storm will affect every aspect of society, economy and culture, unless, to use a Costello phrase, we can fireproof ourselves. We also need to do what we can to diminish the effects of Greenhouse—arguing for an increase in our own emissions, and, having obtained it, refusing to ratify even that watered-down agreement, is not a promising start.

Beware of the demand that the profit motive be the only motive applied to human affairs. The profit motive applied to the environment will destroy the environment. This is not just a matter for esoteric debate: the environment is where we live. It is devastating that after a long slow trudge up the hill, to where the public, and politicians, were beginning to understand the need for conservation as a matter of survival, the debate has been sent back to the bottom of the hill again. We must once again start a slow climb, up the well-worn groove, but each time we do the clock has ticked on further, and there is less time to get to the top before disaster strikes.

There is a clear distinction between the Horton and the Tindale (and Hallam, Jones and Flannery) views of the past, and a clear choice about their implications for the future management of the Australian environment. We need to gamble one way or the other. Whether I am right or wrong, action based on my view of the past provides a good chance of positive future outcomes for the Australian environment; action based on the alternative view does not. Can we afford that gamble?

The idea that Aborigines were on a kind of treadmill, endlessly pedalling faster in order to compensate for the terrible environmental damage they caused 50 000 years ago, is one that should be consigned to the dustbin of history. Consigned to the same dustbin should be the proposition that the human race should mindlessly keep increasing its numbers and, as a result, be forced to pedal the development wheel faster and faster in order to keep up.

In the long term, reducing the population of the world as a whole, and of individual countries, is the only way to ensure a future for the environment and the human race. Aborigines teach us that maintaining an equilibrium with your environment is both a possible and a desirable state.

Clearly we can't go back to the past. Not only does the current population level mean that food production must continue and increase, but the damage that has been caused means that new remedies have to be sought.

There has recently been the encouraging announcement that battery hens will be outlawed after 2012 in Europe. On top of the problems with mad cow disease in Britain and poultry in Belgium, and something of a revolt against genetically modified food, this may be a sign that there is a change in the air. Having spent a century in which farms become more and more removed from nature, less and less diverse, and more and more like factories, it is time to start reversing the trend. Simplification may lead to short-term gains in efficiency, but at the cost of long-term profitability of both individual and country.

Some farmers do things in a particular way because that is the way their fathers and grandfathers did them, the way it has always been done in the district: growing a single crop year after year with increasing soil problems, or running a single breed of animals in the largest numbers possible. Others are able to break away from their history, find new crops and breeds, diversify their production, replant the trees that were ripped out by their great-grandfathers. Such history conquerors will try not only to conserve the biodiversity of their farm but to increase it.

Away from farms, communities should take responsibility for the land in their towns and cities, studying their local histories to see what has changed and what can be returned. Both the old, whose history it is, and the young, whose future it is, need to be involved.

And there need to be the large-scale projects that only governments can undertake. Governments need to change their history of acting for special interest groups and really try to act for the benefit of the whole community, present and future. The Coalition has recently passed legislation that hands environmental matters back to the states, in this as in so much else sending us back to the 1950s. It is crucial that environmental matters are handled at national level, partly because the environment doesn't recognise state borders, but also because state governments have little concern beyond development and short-term gain. Only an Australian goverment is capable of protecting the whole environment, and this latest retrograde move must be reversed as soon as possible.

As I look from my window I see a few dead trees, some ringbarked and left standing last century, some perhaps killed by rising salt in this decade. Good farm-keeping would suggest that these should be knocked down, bulldozed into piles—the plumes of smoke I see rising in the autumn show where many old piles of dead logs are being burnt to produce absolutely clear paddocks. But I leave them, conscious that they form a vital habitat, either standing or blown down in the wind, for many animals. They are the kind of habitat that extensive use of fire would eliminate.

In a nearby district, one where rising smoke in autumn shows where stubble in paddocks is being heartbreakingly burnt—either because people's ancestors have always done it, or because of a mistaken belief that Aborigines did this—people are, amazingly, also being encouraged to leave dead trees standing. The local council wants to save an endangered parrot species.

Each time I drive past the signs that announce this program I am encouraged. There are many signs, blowing in the wind, that the times are changing again. Thinking locally and acting locally is the message from Aborigines as a result of 50 000 years of custodianship of Australia and care for the Australian environment.

Notes

Chapter 1—'Paved with good intentions': Theories on Aborigines and the environment

1 R. Pulleine, 'Presidential Address', *Australian Association for the Advancement of Science* 19, 1998: 294–314.

2 This account is from David Hancock, 'Black Market', *Australian Magazine* 17–18 October 1988: 32–3.

3 The work quoted here is his major work, *The Aboriginal Tribes of Australia*, ANU Press, Canberra, 1974.

4 J. Flood, *Rock Art of Aboriginal Australia*, Penguin, Melbourne, 1983.

5 *Genesis* 1:26, 28.

6 The latest book by Jared Diamond is *Guns, Germs and Steel*, Vintage, London, 1998.

7 The phrase 'intelligent parasitism' seems to have been first used by J. W. Cleland in 1957 (*Mankind* 5:149–62), but the view of Aboriginal–environmental interaction it represents, and the underlying view of the worth of Aboriginal occupation of Australia, is much older. On the other hand, it is also a view still being expressed today, for example by Hugh Morgan.

8 In *The Land*, 29 January 1998, is another frightening story. Peter Daley reports: 'Boggabri district farmer, Geoff Young, has reacted angrily to the restrictions imposed by the State Government

through the new Native Vegetation Conservation Act . . . Instead
of protecting the environment, Mr Young said the legislation would
have the opposite effect. "Anybody who has un-worked land close
to the 10 year mark is going to rip it up, and keep doing it, so
they can use it when they need to. This will have a detrimental
effect on the land in the form of erosion," he said. He said much
of his un-worked land was "useless plains grass" which had no
nutritional benefit to stock.' So much for farmers as 'conservationists'. One might suspect that if the Coalition returns to government
in New South Wales, even such minimally protective legislation
will be repealed and it will be again open slather for environmental
destruction.

9   Rhys Jones has published extensively on a range of issues in
Australian prehistory, the work being quoted here is the seminal
'Fire-Stick Farming', in *Australian Natural History* 16, 1969: 224–8.

10 I criticised the theory strongly, along with a number of other
archaeologists, when Flannery first presented it in 1990 in an
academic journal. See *Archaeology in Oceania* 25: 45–67. There has
been little recent *published* criticism, an honourable exception being
Lesley Head, 'Meganesian barbecue', *Meanjin* 54(4), 1995.

Chapter 2—'An unchanging people in an unchanging land': Archaeology and the past

An earlier version of the first part of this chapter appeared as
D. Horton, 'Archaeology in Australia', *Hemisphere* 28, 1984: 365–70.

1   R. Jones, 'The Neolithic, Palaeolithic and the hunting gardeners',
in R. Suggate and M. Cresswell (eds), *Quaternary Studies*, Royal
Society of New Zealand, Wellington, 1975, pp. 21–34.

2   Ibid.

3   See D. Horton, *The Encyclopaedia of Aboriginal Australia*, Aboriginal
Studies Press, Canberra, 1994.

4   See C. Tatz, 'Aboriginality as Civilisation', *Quarterly* 52: 352–62.

5   John Cobley, *Sydney Cove 1788*, Hodder & Stoughton, London,
1962.

6   See David Horton, *Recovering the Tracks*, Aboriginal Studies Press,
Canberra, 1991.

7   Kow Swamp is in northern Victoria, and was a burial ground
excavated in the 1960s. See A. Thorne and P. Macumber,

'Discoveries of Late Pleistocene Man at Kow Swamp, Australia', *Nature* 238, 1972: 316–19.

8   The DNA experiment was publicised in the media but appears not to have been published academically yet. Dr Bryan Sykes of Oxford University and Dr Chris Stringer of the British Museum were involved in the project.

9   There have been a couple of curious exceptions to these general rules. First is the Mormons, mining the past in order to construct genealogies so that more and more of one's ancestors can be named and then saved. The second is the kind of 'archaeology' that focuses on trying to demonstrate the truth behind some Bible story. One of the most bizarre of such attempts has been the search for 'Noah's Ark', excavating for a metaphor.

Chapter 3—'A slow strangulation of the mind?': Eating fish is wrong

An earlier and more extended version of the argument in this chapter is in D. Horton, 'Tasmanian Adaptation', *Mankind*, 12, 1979: 28–34.

1   'The Tasmanian Paradox' in R. Wright (ed.), *Stone Tools as Cultural Markers*, AIAS, Canberra, 1977, p. 203.

2   R. Jones, 'A speculative archaeological sequence for north-west Tasmania', *Records Queen Victoria Museum* 25, Launceston, 1966. See also his PhD thesis, Rocky Cape and the problem of the Tasmanians, University of Sydney, 1971.

3   Jones, 'The Tasmanian Paradox', p. 200.

4   H. Lourandos, 'Aboriginal spatial organisation and population', *Archaeology and Physical Anthropology in Oceania*, 12, 1978: 202–25.

5   Jones, 'The Tasmanian Paradox', pp. 198–9.

6   D. Horton, *Aboriginal Australia* (map), Aboriginal Studies Press, Canberra, 1996.

7   Rhys Jones, Rocky Cape and the problem of the Tasmanians, PhD thesis, University of Sydney, 1971.

8   R. Jones, 'Tasmania: Aquatic machines and offshore islands', in G. Sieveking, I. Longworth and K. Wilson (eds), *Problems in Economic and Social Archaeology*, Duckworth, London, 1976, pp. 235–63.

9   R. Jones, 'Why did the Tasmanians stop eating fish?', in R. Gould (ed.), *Explorations in Ethnoarchaeology*, University of New Mexico Press, Santa Fe, 1978, pp. 11–47.

10 R. Vanderwal and D. Horton, 'Coastal Southwest Tasmania', *Terra Australis*, 9, ANU, Canberra, 1984.

11 G. Robinson in N. Plomley (ed.), *Friendly Mission*, Tasmanian Historical Research Association, Hobart, 1966. All the observations come from Robinson's time in the south-west in 1830, the dates being 24 March, 25 March, 26 March, 1 June, 3 November, 23 November, 5 December, 25 December.

12 J. Milligan, *Vocabulary of the Dialects of some of the Aboriginal Tribes of Tasmania*, Govt Printer, Hobart, 1890, p. 14.

13 G. Lloyd, *Thirty three Years in Tasmania and Victoria*, Houlston & Wright, London, 1862, p. 52.

Chapter 4—'A people so inclined': To farm or not to farm

Earlier versions of the first part of this chapter appeared in *Biology: The Common Threads*, 2 vols, Academy of Science, Canberra, 1990, 1991.

1 Quoted in J. Beaglehole (ed.), *Journal of the Voyage of the Endeavour 1768–71*, Cambridge University Press, Cambridge, 1955, p. 397.

2 J. Flood, *Archaeology of the Dreamtime*, Collins, Sydney, 1983, pp. 222–34.

3 P. White, *Before the White Man*, Reader's Digest, Sydney, 1974, p. 31.

4 Flood, *Archaeology of the Dreamtime*, pp. 222–34.

5 Using this phrase is just an excuse to quote the marvellous analysis by Watkin Tench (in 1788): 'Longevity, I think, is seldom attained by them. Unceasing agitation wears out the animal frame, and is unfriendly to length of days. But, it may be said, the American Indian, in his undebauched state, lives to an advanced period. True; but he has his seasons of repose: he reaps his little harvest of maize, and continues in idleness while it lasts; he kills the roe-buck or the moose-deer, which maintains him and his family for many days, during which cessation the muscles regain their spring, and fit him for fresh toils. Whereas every sun awakens the native of New South Wales to a renewal of labour, to provide subsistence for the present day.' In Watkin Tench, *Sydney's First Four Years*, Angus & Robertson, Sydney, 1961, p. 48.

6 A number of authors, including John Beaton and Rhys Jones, have commented on this.

Chapter 5—'Opened up a landscape': Firestick farming and the control burners

1 J. W. Cleland, 'Our natives and the vegetation of southern Australia', *Mankind*, 5, 19, 1957: 149–62.

2 N. Tindale, 'Ecology of primitive Aboriginal man in Australia', *Biogeography and Ecology in Australia*, Monographiae Biologicae 8, Junk, The Hague, 1959, pp. 36–51.

3 By C. Dortch in C. E. Dortch and B. G. Muir, 'Long Range Sightings of Bush Fires as a Possible Incentive for Pleistocene Voyages to Greater Australia', *Western Australian Naturalist*, 14, 1980: 194–8.

4 R. Jones, 'The geographical background to the arrival of man in Australia and Tasmania', *Archaeology and Physical Anthropology in Oceania*, 3, 1968: 186–215.

5 S. Hallam, *Fire and Hearth*, AIAS, Canberra, 1975, p. 71.

6 John Cobley, *Sydney Cove 1788*, Hodder & Stoughton, London, 1962.

7 I challenge everyone to read the early accounts with an open mind, rather than reading them with the view to looking for evidence of firestick farming. It is absolutely clear that nowhere in Australia is the description of the early observations of the bush observed consistent with a landscape moulded by 'firestick farming' in the kind of model suggested by Jones and Flannery. J. Benson and P. Redpath ('The nature of pre-European native vegetation in southeastern Australia', *Cunninghamia*, 5, 1997: 285–328) have made an independent analysis of the evidence specifically from New South Wales and reached the same conclusion I reached in general terms twenty years ago. It was a conclusion also reached twenty years ago by I. McBryde and P. Nicholson in 'Aboriginal man and the land in south-western Australia', *Studies in Western Australian History*, 3, 1978: 38–42. They noted that early accounts of Aboriginal use of fire should be treated cautiously.

8 J. Birdsell, 'Some population problems involving Pleistocene man', *Cold Spring Harbour Symposium on Quantitative Biology*, 1957, 22: 47–68.

9 P. Kershaw, 'Evidence for vegetation and climatic change in the Quaternary', in R. Henderson and R. Stephenson (eds), *The Geology and Geophysics of North-eastern Australia*, Geological Society of Australia, Brisbane, 1980, pp. 398–402.

10 G. Singh, P. Kershaw and R. Clark, 'Quaternary vegetation and

fire history in Australia', in A. Gill (ed.), *Fire and the Australian Biota*, Academy of Science, Canberra, 1981, pp. 23–54.

11 Robin Clark, ibid.

12 Clark's is a notable early exception to the uncritical acceptance of the pollen record supporting the firestick farming theory. See 'The prehistory of bushfires', in P. Stanbury, *Bushfire!*, Macleay Museum, University of Sydney, 1981.

Chapter 6—'The extinction of such pachyderms': The great megafauna debate

The arguments in the first part of this chapter were originally presented in an earlier form in D. Horton, 'The great megafaunal extinction debate 1879–1979', *The Artefact*, 4, 1979: 11–25.

1 Letter to R. Owen, 10 July 1844 in M. Aurousseau (ed.), *The Letters of Ludwig Leichhardt*, Hakluyt Society, Cambridge, 1968.

2 M. Ridley, *The Origins of Virtue*, The Softback Preview, England, 1997, p. 218.

3 R. Jones, 'The Neolithic, Palaeolithic and the hunting gardeners', in R. Suggate and M. Cresswell (eds), *Quaternary Studies*, Royal Society of New Zealand, Wellington, 1975, pp. 21–34.

4 W. Anderson, 'On the post-Tertiary ossiferous clays near Myall Creek, Bingara', *Records of the Geological Society New South Wales*, 1, 1890: 116–26.

5 R. Tate, 'President's Address', *Transactions, Proceedings and Reports of the Philosophical Society of Adelaide*, 1879, pp. 39–71.

6 C. Wilkinson, 'President's Address', *Proceedings of the Linnean Society of New South Wales*, 9, 1885: 1207–41.

7 Ibid.

8 E. Stirling, 'The physical features of Lake Calabonna', *Memoirs of the Royal Society of South Australia*, 1, 1900: 1–15.

9 As Ann Moyal has splendidly documented. See, for example, her *Scientists in Nineteenth Century Australia*, Cassell, Melbourne, 1976; A. Mozley, 'Evolution and the climate of opinion in Australia 1840–1876', *Victorian Studies*, 10, 1967: 411–30; R. Macleod, 'Evolutionism and Richard Owen 1830–1868', *Isis*, 56, 1965: 259–80.

10 R. Owen, 'On the discovery of the remains of a mastodontoid pachyderm in Australia', *Annals & Magazine of Natural History* 11, 1843: 11.

11 The Owen sequence is found in 'On the discovery of the remains of a mastodontoid pachyderm in Australia', *Annals and Magazine*

Natural History, 11, 1843: 7–12; 'On the fossil mammals of Australia. Part 3 Diprotodon australis', *Royal Society of London Philosophical Transactions*, 160, 1870: 519–78; *Researches on the Fossil Remains of the Extinct Mammals of Australia*, Erxleben, London, 1877; 'Extinct animals of the colonies of Great Britain', *Proceedings of the Royal Colonial Institute*, 10, 1879: 267–97.

12 F. von Mueller, letter to Owen, 24 August 1861, quoted by A. Moyal in *Scientists in Nineteenth Century Australia*.

13 W. Macleay, 'Inaugural Address', *Proceedings of the Linnean Society of New South Wales*, 1, 1876: 83–96.

14 P. Martin and H. Wright, *Pleistocene Extinctions*, Yale University Press, New Haven, 1967. A later related work was P. Martin and R. Klein, *Quaternary Extinctions*, University of Arizona Press, Tucson, 1984.

15 See R. Gillespie, D. Horton, P. Ladd, P. Macumber, T. Rich, R. Thorne, R. Wright, 'Lancefield and the extinction of the Australian megafauna', *Science*, 200, 1978: 1044–8.

16 J. Furby, Megafauna under the microscope, PhD thesis, University of NSW, 1995; J. Furby, R. Fulagar, P. Dodson, I. Prosser, 'The Cuddie Springs bone bed revisited', in M. Smith, M. Spriggs, B. Fankhauser, *Sahul in Review*, ANU, Canberra, 1991, pp. 204–10.

17 R. Jones and J. Bowler, 'Struggle for the savanna', in R. Jones (ed.), *Northern Australia: Options and implications*, ANU, Canberra, 1980, pp. 3–31.

18 G. Nanson, D. Price and S. Short, 'Wetting and drying of Australia over the past 200ka', *Geology*, 20, 1992: 791–4; P. Hesse, 'The record of continental dust from Australia in Tasman Sea sediments', *Quaternary Science Reviews*, 13, 1994: 257–72.

19 See Stephen Jay Gould, *Life's Grandeur*, Jonathan Cape, London, 1996, for an excellent discussion of the 'Drunkard's Walk'.

20 R. Gould in *Living Archaeology*, Cambridge University Press, Cambridge, 1980, pp. 53–60.

21 E.g. J. Bowler, G. Hope, J. Jennings, G. Singh, D. Walker, 'Late Quaternary climates of Australia and New Guinea', *Quaternary Research*, 67, 1976: 359–4; Jones and Bowler, 'Struggle for the savanna'.

22 E.g. J. Hope and G. Hope, 'Palaeoenvironments for man in New Guinea', in R. Kirk and A. Thorne, *The Origins of the Australians*, AIAS, Canberra, 1976, pp. 29–54. Tim Flannery also accepts this view.

23 See for example A. Anderson, 'Faunal depletion and subsistence change in the early prehistory of southern New Zealand', *Archaeology in Oceania*, 18, 1983: 1–10.

184 THE PURE STATE OF NATURE

24 This is an example of even an arch-sceptic like this author falling for certainty. I had accepted the view that the evidence for asteroid impact was good, putting my faith in geochemists and the like, but lately this interpretation is again being questioned, and the dinosaurs too may have an end more like a whimper than a bang. The only certain thing is that those pesky humans, millions of years away from existing, can't be blamed. If the Australian megafauna had died out, say, 20 million years ago, how, one has to ask, would this have been explained by Rhys Jones and Tim Flannery.

Chapter 7—'Most enlightened conservationists'

1   J. Wright, 'Whose country is it anyway?', *Habitat*, 11, 1983: 26–7.

2   R. Kimber, 'Beginnings of farming?', *Mankind*, 10, 1976: 142–50.

3   R. Kimber, 'Reserve use and management in Central Australia', *Australian Aboriginal Studies*, 2, 1984: 12–23.

4   R. Jones, 'The Neolithic, Palaeolithic and the hunting gardeners', in R. Suggate and M. Cresswell (eds), *Quaternary Studies*, Royal Society of New Zealand, Wellington, 1975, pp. 21–34.

5   R. Hynes and A. Chase, 'Plants, Sites and Domiculture', *Archaeology in Oceania*, 17, 1982: 38–50.

6   J. O'Connell and K. Hawkes, 'Alyawarra plant use and optimal foraging theory', in B. Winterhalder and E. Smith (eds), *Hunter-gatherer Foraging Strategies*, University of Chicago Press, Chicago, 1981, pp. 99–125.

7   Ibid.

8   B. Meehan, *Shell Bed to Shell Midden*, AIAS, Canberra, 1982.

9   One of the many serious problems with the Flannery thesis is his belief that desert Aborigines are typical of all Aborigines in their behaviour, for example their 'nomadism'. In fact the reverse is true. Aborigines in the rest of Australia were largely sedentary and had abundant resources and ready access to water.

10 R. Jones, 'Hunters in the Australian Coastal Savanna', in D. Harris (ed.), *Human Ecology in Savanna Environments*, Academic Press, London, 1980, pp. 107–46.

11 A. Newsome, 'The eco-mythology of the red kangaroo in central Australia', *Mankind*, 12, 1980: 327–33.

12 D. Bennett, 'Some aspects of Aboriginal and non-Aboriginal notions of responsibility to non-human animals', *Australian Aboriginal Studies*, 1983/2: 19–24.

13 Jones, 'Hunters in the Australian Coastal Savanna'.

14 D. Rose, 'The saga of Captain Cook', *Australian Aboriginal Studies*,
    1984/2: 24–39
15 Bennett, 'Some aspects of Aboriginal and non-Aboriginal notions
    of responsibility'.
16 Ibid.

Chapter 8—Convict's dilemma

1   M. Ridley, *The Origins of Virtue*, The Softback Preview, England,
    1997, p. 224.
2   D. Rose, 'The saga of Captain Cook', *Australian Aboriginal Studies*,
    1984/2: 24–39.
3   His name was Muta. Quoted by W. E. H. Stanner in *White Man
    Got No Dreaming*, ANU Press, Canberra, 1967, p. 24. In full: 'White
    man got no dreaming/Him go 'nother way/White man, him go
    different/Him got road belong himself.'
4   See the excellent discussion of this topic in Steve Jones, *In the
    Blood*, Harper Collins, London, 1996. Jones points out that almost
    everyone in the Western world is descended from the Emperor
    Nero. He goes on to say 'everyone is related to everyone else'.
5   'He killed his first duck this year . . . He'd get out here, shoot
    the little birds right around the house'—Norah Golden, grand-
    mother of 11-year-old Andrew Golden after he had shot dead four
    children and a teacher in Arkansas. The report appeared in the
    Canberra Times, 27 March 1998.
6   H. Finlayson, 'Observations on the South Australian members of
    the sub-genus Wallabia', *Transactions of the Royal Society of South
    Australia*, 51, 1927: 363–77. This work should be compulsory
    reading for those proposing commercial exploitation of Australia's
    fauna as a 'conservation' measure.
7   See discussion by J. Golson in 'Australian Aboriginal food plants',
    in D. J. Mulvaney and J. Golson, *Aboriginal Man and Environment
    in Australia*, ANU Press, Canberra, 1971, pp. 196–23.

Chapter 9—Ghosts

1   W.E.H. Stanner, *White Man Got No Dreaming*, ANU Press, Can-
    berra, 1967, p. 34.
2   ABC Television's 'Australian Story', 29 April 1999.

Index

Aboriginal culture/civilisation viii, 7, 22, 28, 31, 68, 128
  diversity of 27, 43
  non-Aboriginal perceptions of viii–ix, 2, 3, 22
  see also Dreaming/Dreamtime; religious beliefs; Tasmanian Aborigines
Aborigines
  colonists' perceptions of 3, 14, 27
  and evolutionary theory 8, 9, 11
  land ownership, *see* native title
  origins of 11–12, 25–6, 36, 38, 72–3, 75
  perception of time/the past 12, 31–4, 36–7
  period of occupation of Australia 8, 11, 12, 27, 32, 34–5, 37–8, 75, 87, 108, 110, 114, 170
  stolen generations 165
  Tasmanian, *see* Tasmanian Aborigines
  viewed as a doomed people 39–40, 51
Aborigines and the environment Chapter 1, 26, 29, 34, 51, 57, 103, 140, 142–3, 148–9, 153, 154, 155–6, 157, 162, 163, 171, *see also* conservation; fire; firestick farming theories; megafauna
Adams, Phillip 8
agriculture, *see* farming/agriculture
Alyawarra people 133–4
Anbarra people 134
ancestors, Aboriginal 11
Anderson, William 104
animal products 67–8